T0349919

NEAR-EQUILIBRIUM
TRANSPORT

Fundamentals and Applications

Lessons from Nanoscience: A Lecture Note Series

ISSN: 2301-3354

Series Editors: Mark Lundstrom and Supriyo Datta
(Purdue University, USA)

"Lessons from Nanoscience" aims to present new viewpoints that help understand, integrate, and apply recent developments in nanoscience while also using them to re-think old and familiar subjects. Some of these viewpoints may not yet be in final form, but we hope this series will provide a forum for them to evolve and develop into the textbooks of tomorrow that train and guide our students and young researchers as they turn nanoscience into nanotechnology. To help communicate across disciplines, the series aims to be accessible to anyone with a bachelor's degree in science or engineering.

More information on the series as well as additional resources for each volume can be found at: http://nanohub.org/topics/LessonsfromNanoscience

Published:

Vol. 1   Lessons from Nanoelectronics: A New Perspective on Transport
by Supriyo Datta

Vol. 2   Near-Equilibrium Transport: Fundamentals and Applications
by Mark Lundstrom and Changwook Jeong

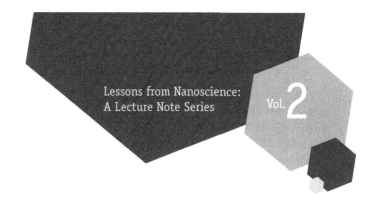

Lessons from Nanoscience:
A Lecture Note Series

Vol. 2

NEAR-EQUILIBRIUM TRANSPORT

Fundamentals and Applications

Mark Lundstrom
Changwook Jeong

Purdue University, USA

 **World Scientific**

NEW JERSEY • LONDON • SINGAPORE • BEIJING • SHANGHAI • HONG KONG • TAIPEI • CHENNAI

Published by

World Scientific Publishing Co. Pte. Ltd.

5 Toh Tuck Link, Singapore 596224

USA office: 27 Warren Street, Suite 401-402, Hackensack, NJ 07601

UK office: 57 Shelton Street, Covent Garden, London WC2H 9HE

British Library Cataloguing-in-Publication Data
A catalogue record for this book is available from the British Library.

Lessons from Nanoscience: A Lecture Note Series — Vol. 2
NEAR-EQUILIBRIUM TRANSPORT
Fundamentals and Applications

Copyright © 2013 by World Scientific Publishing Co. Pte. Ltd.

ISBN 978-981-4327-78-7
ISBN 978-981-4355-80-3 (pbk)

Printed in Singapore by Mainland Press Pte Ltd.

To

Cason Kossuth Lundstrom
and
Sunyoung, Hyebin, and Seyeon.

Preface

Engineers and scientists working on electronic materials and devices need a working knowledge of "near-equilibrium" (also called "linear" or "low-field") transport. By "working knowledge" we mean understanding how to use theory in practice. Measurements of resistivity, conductivity, mobility, thermoelectric parameters as well as Hall effect measurements are commonly used to characterize electronic materials. Thermoelectric effects are the basis for important devices, and devices like transistors, which operate far from equilibrium, invariably contain low-field regions (e.g. the source and drain) that can limit device performance. These lectures are an introduction to near-equilibrium carrier transport using a novel, bottom up approach as developed by my colleague, Supriyo Datta and presented in Vol. 1 of this series [1]. Although written by two electrical engineers, it is our hope that these lectures are also accessible to students in physics, materials science, chemistry and other fields. Only a very basic understanding of solid-state physics, semiconductors, and electronic devices is assumed. Our notation follows standard practice in electrical engineering. For example, the symbol, "q", is used to denote the magnitude of the charge on an electron and the term, Fermi level (E_F), is used for the chemical potential in the contacts.

The topic of near-equilibrium transport is easy to either over-simplify or to encumber by mathematical complexity that obscures the underlying physics. For example, ballistic transport is usually treated differently than diffusive transport, and this separation obscures the underlying unity and simplicity of the field. These lectures provide a different perspective on traditional concepts in electron transport in semiconductors and metals as well as a unified way to handle macroscale, microscale, and nanoscale devices. A short introduction to the Boltzmann Transport Equation (BTE), which is commonly used to describe near-equilibrium transport, is also included and related to the approach used here. Throughout the lectures, concepts are illustrated with examples. For the most part, electron transport with a simple, parabolic energy band structure is assumed, but the approach is much more general. A short chapter shows, for example, how the same approach can be applied to the transport of heat by phonons, and to illustrate how the theory is applied to new problems. The lectures conclude with a case study – near-equilibrium transport in graphene.

It should, of course, be understood that this short set of lectures is only a starting point. The lectures seek to convey the essence of the subject and prepare students to learn. The additional topics needed to address specific research, development, and engineering problems on their own. Online versions of these lectures are available, along with an extensive set of additional resources for self-learners [2]. In the spirit of the *Lessons from Nanoscience* Lecture Note Series, these notes are presented in a still-evolving form, but we hope that readers find them a useful introduction to a topic in electronic materials and devices that continues to be relevant and interesting at the nanoscale.

Mark Lundstrom
Changwook Jeong
Purdue University
June 18, 2012

[1] Supriyo Datta, *Lessons from Nanoelectronics: A new approach to transport theory*, Vol.1 in *Lessons from Nanoscience: A Lecture Notes Series*, World Scientific Publishing Company, Singapore, 2011.

[2] M. Lundstrom, S. Datta, and M.A. Alam, "Lessons from Nanoscience: A Lecture Note Series", http://nanohub.org/topics/LessonsfromNanoscience, 2011.

Acknowledgments

Thanks to World Scientific Publishing Corporation and our series editor, Zvi Ruder, for their support in launching this new lecture notes series. Special thanks to the U.S. National Science Foundation, the Intel Foundation, and Purdue University for their support of the Network for Computational Nanotechnology's "Electronics from the Bottom Up" initiative, which laid the foundation for this series.

Students at Purdue University, Norfolk State University, Dalian University of Technology, the University of Pisa, and attendees of the 2011 NCN Summer School served as sounding boards and proof-readers for these notes. Their comments and suggestions are appreciated as is the help of students who taught one of us (Lundstrom) enough LaTex to get the job done. Dr. Jesse Maassen's help with the final proof-reading is appreciated, and special thanks go to Dr. Raseong Kim, whose initial work was the genesis for these notes and who supplied the compilation of thermoelectric coefficients presented in the appendix. Finally, we acknowledge many discussions with Professor Supriyo Datta, whose ideas and thinking have strongly influenced this work.

Contents

List of Figures

Lecture 1

Overview

Contents

1.1   Introduction

This short set of lectures is about how electrons in semiconductors and metals flow in response to driving forces such as applied voltages and differences in temperature. The simplest description of transport is the famous Ohm's Law (Georg Ohm, 1927),

$$I = V/R = GV \,, \tag{1.1}$$

which states that the current through a conductor is proportional to the voltage across it. One goal of these lectures is to develop an understanding of why and under what conditions the current-voltage characteristic is linear and to understand how the resistance is related to the material properties of the resistor. Before launching into the lectures, let's spend a few minutes discussing what the following lectures are all about.

1.2   Diffusive electron transport

The transport of charge carriers such as electrons is a rich and deep field of physics. While we won't be delving into the underlying physics in great detail, it will be necessary to have a firm grasp of some fundamentals. Consider Fig. 1.1, which illustrates diffusive electron transport in a simple resistor (made with an n-type semiconductor for which the current is carried by electrons in the conduction band).

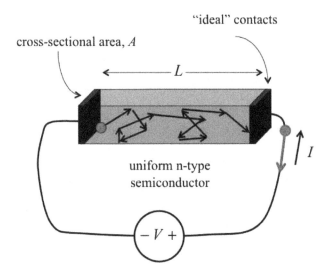

Fig. 1.1.   Illustration of diffusive electron transport in an n-type semiconductor under bias.

Because the positive voltage on the right contact attracts electrons, they tend to flow from left to right, but it is a random walk during which electrons frequently scatter from defects, impurities, etc. In the traditional approach, we say that electrons feel a force due to the electric field,

$$F_e = -q\mathcal{E}_x = qV/L\,. \tag{1.2}$$

The electric field accelerates electrons, but scattering produces an opposing force, so the result is that in the presence of an electric field, electrons drift at a steady-state velocity of

$$\upsilon_d = -\mu_n\mathcal{E}_x = \mu_n V/L\,, \tag{1.3}$$

where μ_n is the mobility, a material-dependent parameter.

We obtain the current by noting that it is proportional to the charge on an electron, q, the density of electrons per unit volume, n, the cross-sectional area, A, and to the drift velocity, v_d. The current can be written as

$$I = nq\mu_n \frac{A}{L} V = GV \, , \qquad (1.4)$$

where

$$G = nq\mu_n \frac{A}{L} = \sigma_n \frac{A}{L} \, , \qquad (1.5)$$

with G being the conductance in Siemens (S = 1/Ohms), and σ_n the conductivity in S/m (1/Ohm-m). Equation (1.4) is a classic result that we shall try to understand more deeply.

Figure 1.2 is a sketch of what a measured I–V characteristic might look like for a semiconductor like silicon. We see that there is a region for which the current varies linearly with voltage. (This is the regime of near-equilibrium, linear, or low-field transport that we shall be concerned with. Under high bias, the current becomes a non-linear function of voltage (and may even be non-monotonic, as in semiconductors like GaAs). A proper discussion of high-field transport would require another set of lecture notes. The interested reader can consult Chapter 7 of Lundstrom [1].

Figure 1.3 illustrates the kind of problem that engineers and applied scientists are increasingly dealing with — an extremely short conductor, in this case a small molecule. The resistance of devices like this can be

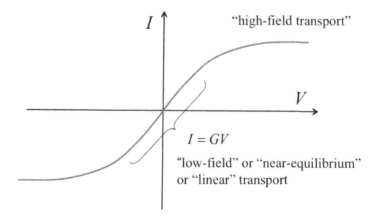

Fig. 1.2. Illustration of a typical current vs. voltage characteristic for a semiconductor like silicon.

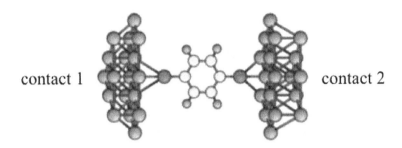

Fig. 1.3.   Illustration of a small organic molecule (phenyl dithiol) attached to two gold contacts. The *I–V* characteristics of small molecules can now be measured experimentally. See, for example, L. Venkataraman, J. E. Klare, C. Nuckolls, M. S. Hybertsen, and M. L. Steigerwald, "Dependence of single-molecule junction conductance on molecular conformation", *Nature*, **442**, 904-907, 2006.

measured, and we need a theory to understand the measured resistance. Equation (1.4), seems ill-suited to this problem, but a general transport theory should be able to treat both large conductors and very small ones. When we develop this theory, we will find some surprises. For example, we'll find that there is an upper limit to the conductance, no matter how short the conductor is and that conductance comes in quantized units. These facts are well-known from research on mesoscopic physics (e.g. see Datta [2]) but they have now become important in device research and engineering.

1.3   Types of electron transport

Electronic transport is a rich and complex field. Some important types of transport are:

(1) near-equilibrium transport
(2) high-field (or hot carrier) transport
(3) non-local transport in small devices
(4) quantum transport
(5) transport in random/disordered/nanostructured materials
(6) phonon transport

Near-equilibrium transport is diffusive in conductors that are many mean-free-paths long, but it is ballistic when the conductor is much shorter than a mean-free-path. We shall discuss both cases. High field transport in bulk semiconductors is also diffusive, but since the carriers gain significant kinetic energy from the high electric field, they are more energetic (hotter) than the lattice. Under high applied biases, the current is a nonlinear function of the applied voltage, and Ohm's Law no longer holds. (See Chapter 7 of Lundstrom [1] for a discussion of high field transport.)

In modern semiconductor devices, modest applied biases (e.g. 1 V) can produce large electric fields across the short, active regions. Under these conditions, carrier transport becomes a nonlocal function of the electric field, and interesting effects such as velocity overshoot can occur. (See Chapter 8 of Lundstrom [1] for a discussion of these effects.) Finally, as devices get very short and the potential changes rapidly on the scale of the electron's wavelength, effects like quantum mechanical reflection and tunneling become important. For an introduction to the field of quantum tranport, see Datta [3].

Another important class of problems has to do with transport in various kinds of disordered materials. Much of traditional transport theory assumes a periodic crystal lattice and makes use of concepts like crystal momentum and Brillouin zones. In these lectures, we will restrict our attention to this class of problems. Amorphous materials, however, do not possess long range order; some interesting new features occur, but many of the essential aspects of scattering are similar (e.g. see Mott and Davis [4]). Polycrystalline materials consist of single crystal grains separated by grain boundaries. The statistical distributions of grain sizes, orientations, and grain boundary properties complicate the description of electron transport. Indeed, much of the promise of nanotechnology lies in the hope that artificially structuring matter at the nanoscale will provide properties not found in nature. We believe that the approach used in these lectures will prove useful for this class of problems as well, but this is a topic of current research, and the lecture notes volume will have to wait.

Finally, we should mention that although our attention in these lectures is on electron transport in semiconductors and metals. Phonon transport is also important, and many of the concepts developed to describe electron transport can just as well describe phonon transport. The calculation of the thermal conductivity is much like the calculation of the electrical conductivity. Just as for electrons, there can be near-equilibrium, diffusive, ballistic, far-from-equilibrium, and quantum (or wave-like) transport of phonons in crystalline or disordered materials. In modern integrated circuits, power dissipation from electron transport heats the lattice and generates phonons. Understanding how to manage this problem requires us to understand both electron and phonon transport. The performance of thermoelectric devices used for electronic cooling and for electrical power generation from thermal gradients is controlled by both electron and phonon transport. Although our focus in these lectures is on electron transport, a brief discussion of phonon transport is also included.

1.4   Why study near-equilibrium transport?

Given that near-equilibrium transport in crystalline materials is only a small subset of carrier transport, one might question the need to devote an entire volume to the topic. There are some good reasons. First, near-equilibrium transport is the foundation for understanding transport in general. Concepts introduced in the study of near-equilibrium are often extended to treat more complicated problems, and near-equilibrium transport provides a reference point for comparison when we analyze transport in more complex situations. Second, near-equilibrium transport measurements are extensively used to characterize electronic materials and to understand the properties of new materials. Finally, near-equilibrium transport controls or strongly influences the performance of most electronic devices.

There is a very large number of books that discuss low-field transport from a traditional perspective — typically using the Boltzmann Transport Equation (e.g. see Refs. [5-7]). In this volume, I use a new approach that I believe is more physically transparent, mathematically more simple, and that is more broadly applicable.

1.5   About these lectures

The list of lectures presented in this collection is:

Lecture 1: Overview
Lecture 2: General Model for Transport
Lecture 3: Resistance: Ballistic to Diffusive
Lecture 4: Thermoelectric Effects: Physical Approach
Lecture 5: Thermoelectric Effects: Mathematics
Lecture 6: An Introduction to Scattering
Lecture 7: Boltzmann Transport Equation
Lecture 8: Near-equilibrium Transport: Measurements
Lecture 9: Phonon Transport

Lecture 10: Graphene: A Case Study

A brief description of each of these lectures follows.

Lecture 2: General Model for Transport

Datta's model of a nanodevice (a version of the Landauer approach) [8] is introduced as a general way to describe transport in nanodevices — as well as in bulk metals and semiconductors.

Lecture 3: Resistance: Ballistic to Diffusive

The resistance of a ballistic conductor and concepts such as the quantum contact resistance are introduced and discussed. The results are then generalized to treat transport all the way from the ballistic to diffusive regimes. We will show how to treat bulk conductors (electrons free to move in 3D) and will also discuss 2D conductors (electrons free to move in a plane) and 1D conductors (electrons free to move along a wire).

Lecture 4: Thermoelectric Effects: Physical Approach

The effect of temperature gradients on current flow and how electrical currents produce heat currents will be discussed. Coupled equations for the electric and heat currents will be presented, and applications to electronic cooling and the generation of electrical power from thermal gradients will be briefly discussed. In this lecture, we use a physical approach and try to keep the mathematics to a minimum.

Lecture 5: Thermoelectric Effects: Mathematics

Beginning with the general model for transport, we mathematically derive expressions for the four thermoelectric transport coefficients:

 (i) Electrical conductivity
 (ii) Seebeck coefficient (or "thermopower")
(iii) Peltier coefficient
(iv) Electronic heat conductivity

We also discuss the relationship of the coefficients (e.g. the Kelvin relation and the Wiedemann-Franz Law).

Lecture 6: An Introduction to Scattering

In Lectures 1-5, scattering is described by a mean-free-path (MFP) for backscattering. In this lecture, we show how the MFP is related to the time between scattering events and briefly discuss how the scattering time is related to underlying physical processes.

Lecture 7: Boltzmann Transport Equation

Semi-classical carrier transport is traditionally described by the Boltzmann Transport Equation (BTE) (e.g. [1, 5-7]). In this lecture, we present the BTE, show how it is solved, and relate it to the Landauer approach used in these lectures. As an example of the use of the BTE, we derive the conductivity in the presence of an applied B-field.

Lecture 8: Near-equilibrium Transport: Measurements

Measurements of near-equilibrium transport are routinely used to characterize electronic materials. This lecture is a brief introduction to commonly-used techniques such as van der Pauw and Hall effect measurements.

Lecture 9: Phonon Transport

Most of the heat flow in semiconductors is carried by phonons (i.e. quantized lattice vibrations). In the presence of a small temperature gradient, phonon transport is also a problem in near-equilibrium transport, and the techniques developed for electron transport can be readily extended to phonons. This lecture is an introduction to phonon transport. Key similarities and differences between electron and phonon transport are discussed.

Lecture 10: Graphene: A Case Study

In Lectures 1-8 we largely consider applications of near-equilibrium electron transport to traditional materials, such as semiconductors with parabolic energy bands, but the theory is much more general. As an example of how to apply the concepts in these lectures, we discuss near-equilibrium transport in graphene, a material that has recently attracted a lot of attention and was the subject of the 2010 Nobel Prize in Physics.

Appendix: Brief Summary of Key Results

The central ideas conveyed in these notes are easy to grasp, but the notes contain many equations so that the reader can see all the steps in the derivations of key results. To assist the reader in performing computations, the key results are summarized in this short appendix, which includes pointers to specific results in the various lectures. Expressions of the four transport parameters for materials with simple bandstructures are often needed and are also listed in this appendix.

1.6 Summary

My objectives for this collection of lectures are very simple:

(1) To introduce the essentials of near-equilibrium carrier transport using a "bottom up" approach that works at the nanoscale as well as at the macroscale.
(2) To acquaint students with some key results (e.g. the quantum of conductance, common measurement techniques).
(3) To provide a basic foundation upon which students can build as they encounter problems in research and engineering.

Your goal in reading these lecture notes should be to acquire a firm understanding of the fundamental concepts and to develop an ability to apply these fundamentals to real problems. Those interested in developing a deeper understanding of the physics of transport should consult Refs. [2, 3, 8].

1.7 References

For an introduction to high-field transport and to non-local transport in semiconductor devices, see Chapters 7 and 8 in:

[1] Mark Lundstrom, *Fundamentals of Carrier Transport* 2nd *Ed.*, Cambridge Univ. Press, Cambridge, U.K., 2000.

Chapters 1 and 2 in the following book are a good introduction to the so-called Landauer approach.

[2] Supriyo Datta, *Electronic Transport in Mesoscopic Systems*, Cambridge Univ. Press, Cambridge, U.K., 1995.

For an introduction to quantum transport, see:

[3] Supriyo Datta, *Quantum Transport: Atom to transistor*, Cambridge Univ. Press, Cambridge, U.K., 2005.

For a classic introduction to electronic conduction in noncrystalline materials, see:

[4] N.F. Mott and E.A. Davis, *Electronic Processes in Non-Crystalline Materials*, Clarendon Press, Oxford, U.K., 1971.

Three classic references on low-field transport are:

[5] J.M. Ziman, *Principles of the Theory of Solids*, Cambridge Univ. Press, Cambridge, U.K., 1964.

[6] A.C. Smith, J. Janak, and R. Adler, *Electronic Conduction in Solids*, McGraw-Hill, New York, N.Y. 1965.

[7] N.W. Ashcroft and N.D. Mermin, *Solid–State Physics*, Saunders College, Philadelphia, PA, 1976.

The conceptual approach used in these lectures is presented in a succinct form by Datta:

[8] Supriyo Datta, *Lessons from Nanoelectronics: A new approach to transport theory*, World Scientific Publishing Company, Singapore, 2011.

A collection of additional resources on carrier transport can be found at:

[9] Mark Lundstrom and Supriyo Datta, "Electronics from the Bottom Up", http://nanohub.org/topics/ElectronicsFromTheBottomUp, 2011.

Hear a lecture on this chapter at:

[10] M. Lundstrom, "General Model for Transport", http://nanohub.org/ topics/LessonsfromNanoscience, 2011.

General Model for Transport

Contents

2.1   Introduction

In this chapter, we introduce a simple, but surprisingly useful model for electron transport. As sketched in Fig. 2.1, we first seek to understand the I–V characteristics of a nanoscale electronic device. The approach is due to R. Landauer in a form developed by Datta [1–3]. As indicated in Fig. 2.1, the contacts play an important role, but we will see that the final result can be generalized to describe transport in the bulk, for which the current is controlled by the properties of the material between the contacts. We shall not attempt to spatially resolve quantities within the device. In practice this can be important, especially for devices under moderate or high bias. Semiconductor devices are often described by the so-called semiconductor equations [4], that make use of the type of bulk transport equation that we shall discuss.

The heart of the device, the channel, is a material that is described by its density-of-states, the DOS, $D(E - U)$, where E is energy, and U a self-consistent electrostatic potential, U. An external gate may be used to

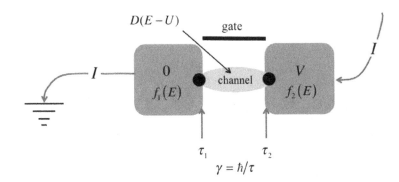

Fig. 2.1.   Illustration of a model nanoscale electronic device. The voltage, V, lowers the Fermi level of contact 2 by an amount, qV.

move the states up and down in energy (as in a transistor), but in these lectures we will assume a two-terminal device and set $U = 0$.

The channel of our device is connected to two ideal contacts, which are assumed to be large regions in which strong scattering maintains near-equilibrium conditions. Accordingly, each contact is described by an equilibrium Fermi function (or occupation number),

$$f_0 = \frac{1}{1 + e^{(E-E_F)/k_B T_L}} \,, \tag{2.1}$$

where E_F is the Fermi level (chemical potential) of the contact, and T_L is the temperature of the lattice, which is also the temperature of the electrons because of the assumed near-equilibrium conditions. Each of the two contacts is in equilibrium, but if a voltage, V, is applied across the device, then $E_{F2} = E_{F1} - qV$.

The connection of the contacts to the channel is described by a characteristic time, τ, which describes how long it takes electrons to get in and out of the device. For a very small device (e.g. a single molecule), τ is controlled by the contact. For longer devices with good contacts, we will see that τ becomes the transit time for electrons to cross the channel. In general, the two connections might be different, so τ_1 and τ_2 may be different. Sometimes it is convenient to express τ in units of energy according to $\gamma = \hbar/\tau$. If the channel is a single molecule, γ has a simple physical interpretation; it represents the "broadening" of the molecular energy levels due to the finite lifetimes of the electrons in a molecular level.

Although this model is very simple, we shall see that it is also very powerful. We shall be concerned with two questions:

(1) How is the electron density in the device related to the Fermi levels in the contacts, to the density of states, and to the characteristic times?
(2) How is the electron current through the device related to the same parameters?

Before we develop the mathematical model, we briefly summarize the key assumptions. For a fuller discussion of these assumptions, see Refs. [1–3].

(1) The channel of the device is described by a band structure, $E(k)$. This assumption is not necessary; for the more general case, see Ref. [1].
(2) The contacts are large with strong inelastic scattering that maintain near-equilibrium conditions.
(3) We assume that electrons feel a self-consistent (mean-field) potential, U, due to the other electrons and the applied biases. (This assumption breaks down for "strongly correlated transport", such as single electron charging.) In practice, we would find the self-consistent potential by solving the Poisson equation. It is important for devices like transistors, but in these notes, we restrict our attention to two-terminal devices and set $U = 0$.
(4) All inelastic scattering takes place in the contacts. Electrons flow from left to right (or right to left) in independent energy channels.
(5) The contacts are reflectionless (absorbing). Electrons that enter the contact from the channel are equilibriated according to the Fermi level of the contact.

Although these assumptions may appear restrictive, we will find that they describe a large class of problems. Having specified the model device, we turn next to the mathematical analysis.

2.2 Mathematical model

To develop the mathematical model, consider first the case where only the first (left) contact is connected to the channel. Contact 1 will seek to fill up the states in the channel according to E_{F1}. Eventually, contact 1 and the channel will be in equilibrium with number of electrons between E and

$E + dE$ given by

$$N'_{01}(E)dE = D(E)dEf_1(E),\qquad(2.2)$$

where $D(E)$ is the density-of-states at energy, E, in the channel and $f_1(E)$ is the equilibrium Fermi function of contact 1. Note that $N'_{01}(E)dE$ is the total number of electrons, not the number density. The density-of-states includes the factor of two for spin degeneracy. We can also write a simple rate equation to describe the process by which equilibrium between the contact and channel is achieved. The rate equation is

$$F_1 = \left.\frac{dN'(E)}{dt}\right|_1 = \frac{N'_{01}(E) - N'(E)}{\tau_1(E)}.\qquad(2.3)$$

According to eqn. (2.3), dN'/dt is positive if the number of the electrons in the channel is less than the equilibrium number and negative if it is more. If the channel is initially empty, the channel fills up until equilibrium is achieved, and if is initially too full of electrons, it empties out until equilibrium with the contact is reached.

On the other hand, if only contact 2 is connected to the channel, a similar set of equations can be developed,

$$N'_{02}(E)dE = D(E)dEf_2(E),\qquad(2.4)$$

$$F_2 = \left.\frac{dN'(E)}{dt}\right|_2 = \frac{N'_{02}(E) - N'(E)}{\tau_2(E)}.\qquad(2.5)$$

In practice, both contacts are connected at the same time and both inject or withdraw electrons from the channel. The total rate of change of the electron number in the device is

$$\left.\frac{dN'(E)}{dt}\right|_{tot} = F_1 + F_2 = \left.\frac{dN'(E)}{dt}\right|_1 + \left.\frac{dN'(E)}{dt}\right|_2.\qquad(2.6)$$

In steady-state, $dN'/dt = 0$, and we can solve for the steady-state number of electrons in the channel as

$$N'(E)dE = \frac{D(E)dE}{2}f_1(E) + \frac{D(E)dE}{2}f_2(E),\qquad(2.7)$$

where we have assumed that $\tau_1 = \tau_2$ and used eqns. (2.2) and (2.4). Finally, we obtain the total, steady-state number of electrons in the channel by integrating over all of the energy channels,

$$N = \int N'(E)dE = \int \left[\frac{D(E)}{2} f_1(E) + \frac{D(E)}{2} f_2(E) \right] dE . \qquad (2.8)$$

Equation (2.8) is the answer to our first question. It gives the number of electrons in the channel of the device in terms of the density-of-states of the channel and the Fermi functions of the two contacts. Finally, a word about notation. The quantity, N' has units of number / energy; it is the differential carrier density, $N'(E) = dN/dE|_E$.

We should note the similarity of eqn. (2.8) to the standard expression for the equilibrium electron number in a semiconductor [4],

$$N_0 = \int D(E)f_0(E)dE . \qquad (2.9)$$

The difference is that eqn. (2.9) refers to the number of electrons in equilibrium whereas eqn. (2.8) describes a device that may be in equilibrium (if $E_{F1} = E_{F2}$) or very far from equilibrium if the Fermi levels are very different.

We should remember that N is the total *number* of electrons in the channel, and $D(E)$ is the total density-of-states, the number of states per unit energy. In 1D, $D \propto L$, the length of the channel. In 2D, $D \propto A$, the area of the channel, and in 3D, $D \propto \Omega$, the volume of the channel. For device work we usually prefer to express the final answers in terms of the electron density (per unit length in 1D, per unit area in 2D, and per unit volume in 3D).

Having answered our first question, how the electron number is related to the properties of the channel and contacts, we now turn to the second question, the steady-state current. When a steady-state current flows, one contact tries to fill up states in the channel and the other tries to empty them. If $E_{F1} > E_{F2}$, contact 1 injects electrons and contact 2 removes them, and vice versa if $E_{F1} < E_{F2}$.

The rates at which electrons enter or leave contacts 1 and 2 are given by eqns. (2.3) and (2.5). In steady state,

$$F_1 + F_2 = 0 . \qquad (2.10)$$

The current is *defined* to be positive when it flows into contact 2, so

$$I' = qF_1 = -qF_2 .\qquad(2.11)$$

Using our earlier results, eqns. (2.2) and (2.4), we find

$$I'(E) = \frac{q}{2\tau(E)}\left(N'_{01} - N'_{02}\right) = \frac{2q}{h}\frac{\gamma(E)}{2}\pi D(E)\left(f_1 - f_2\right) ,\qquad(2.12)$$

where

$$\gamma \equiv \frac{\hbar}{\tau(E)} ,\qquad(2.13)$$

Finally, the total current is found by integrating over all of the energy channels,

$$I = \int I'(E)dE = \frac{2q}{h}\int \gamma(E)\pi\frac{D(E)}{2}\left(f_1 - f_2\right)dE .\qquad(2.14)$$

According to eqn. (2.14), current only flows when the Fermi levels of the two contacts differ. In that case, there is a competition — one contact keeps trying to fill up the channel while the other one keeps trying to empty it.

This concludes the mathematical derivation that answers our two questions about how the steady-state number of electrons and current are related to the properties of the channel and contacts. The key results, eqns. (2.8) and (2.14) are repeated below.

$$\boxed{\begin{aligned} N &= \int \frac{D(E)}{2}\left(f_1 + f_2\right)dE \\ I &= \frac{2q}{h}\int \gamma(E)\pi\frac{D(E)}{2}\left(f_1 - f_2\right)dE . \end{aligned}}\qquad(2.15)$$

The remainder of these lecture notes largely consists of understanding and applying these results.

2.3   Modes

The fact that the current is proportional to $(f_1 - f_2)$ makes sense, and $2q/h$ is a set of fundamental constants that we shall see is important, but what is the product, $\gamma\pi D/2$? It is an important quantity. According to eqn. (2.13),

γ has units of energy. The density-of-states, $D(E)$, has units of $1/$energy. (Recall that we deal with total electron numbers, not electron densities, so the 3D density-of-states here does not have units of $1/$energy-volume, as is customary in semiconductor physics.) Accordingly, we conclude that the product, $\gamma \pi D/2$ is dimensionless. We shall see that it is the number of conducting channels at energy, E.

Figure 2.2 is a sketch of a two-dimensional, ballistic channel.

The total density-of-states is

$$D(E)/A = D_{2D}(E) = g_v \frac{m^*}{\pi \hbar^2} , \tag{2.16}$$

where D_{2D} is the 2D density-of-states per unit area, the number of states per J-m^2. The final result assumes parabolic energy bands with an effective mass of m^* and occupation of a single subband (due to confinement in the vertical direction) with a valley degeneracy of g_v.

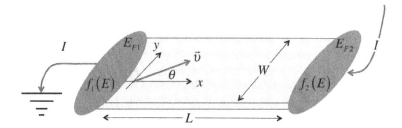

Fig. 2.2. A simple, 2D electronic device with channel width, W and length, L. For the calculation of the average x-directed velocity, ballistic transport is assumed, i.e. the channel is much shorter than a mean-free-path for scattering.

Let's do an "experiment" to determine the characteristic time, τ. From eqns. (2.7) and (2.12), we find

$$\frac{qN'(E)dE}{I'(E)dE} = \frac{\hbar}{\gamma} \frac{(f_1 + f_2)}{(f_1 - f_2)} . \tag{2.17}$$

Now in our experiment we apply a large voltage to contact 2, which makes $E_{F2} \ll E_{F1}$ so $f_2 \ll f_1$, and eqn. (2.17) becomes

$$\frac{qN'(E)dE}{I'(E)dE} = \frac{\text{stored charge}}{\text{current}} = \frac{\hbar}{\gamma} = \tau(E) . \tag{2.18}$$

The number of electrons in the channel is $N'(E) = n'_s(E)WL$, where n_s is the electron density per unit area. The differential current can be written as $I'(E) = qWn'_s(E)\langle v_x^+(E)\rangle$, so from eqn. (2.18), we find

$$\tau(E) = \frac{L}{\langle v_x^+(E)\rangle},\qquad(2.19)$$

which is just the average transit time of carriers across the channel.

To evaluate $\tau(E)$, we need $\langle v_x^+(E)\rangle$, the average velocity in the $+x$ direction. From Fig. 2.2, we see that for ballistic transport, in which electrons travel across the device without changing direction,

$$\langle v_x^+(E)\rangle = v(E)\langle\cos\theta\rangle .\qquad(2.20)$$

A simple calculation gives

$$\langle\cos\theta\rangle = \frac{\int_{-\pi/2}^{\pi/2}\cos\theta d\theta}{\pi} = \frac{2}{\pi},\qquad(2.21)$$

so we find the average ballistic velocity in the $+x$ direction as

$$\langle v_x^+(E)\rangle = \frac{2}{\pi}v = \frac{2}{\pi}\sqrt{\frac{2(E-E_C)}{m^*}},\qquad(2.22)$$

where the final result assumes parabolic energy bands. (We also assumed isotropic conditions, so that $v(E)$ is not a function of θ.) Defining

$$M(E) \equiv \gamma(E)\pi\frac{D(E)}{2}\qquad(2.23)$$

and using $\gamma = \hbar/\tau$ and $D = D_{2D}WL$, we find

$$M(E) = WM_{2D}(E) = W\frac{h}{4}\langle v_x^+(E)\rangle D_{2D}(E).\qquad(2.24)$$

Similar arguments in 1D and 3D yield

$$\boxed{\begin{aligned}
M(E) &= M_{1D}(E) = \frac{h}{4}\langle v_x^+(E)\rangle D_{1D}(E)\\[1em]
M(E) &= WM_{2D}(E) = W\frac{h}{4}\langle v_x^+(E)\rangle D_{2D}(E)\\[1em]
M(E) &= AM_{3D}(E) = A\frac{h}{4}\langle v_x^+(E)\rangle D_{3D}(E).
\end{aligned}}\qquad(2.25)$$

Note that the number of conducting channels at energy, E, is proportional to the width of the conductor in 2D and to the cross-sectional area in 3D.

We now have expressions for the number of channels at energy, E in 1D, 2D, and 3D, but we should try to understand the result. For parabolic energy bands, we can evaluate (2.24) to find

$$WM_{2D}(E) = g_v W \frac{\sqrt{2m^*(E - E_c)}}{\pi\hbar} , \qquad (2.26)$$

where g_v is the valley degeneracy. Parabolic energy bands are described by

$$E(k) = E_C + \frac{\hbar^2 k^2}{2m^*} , \qquad (2.27)$$

which can solved for k to write

$$WM_{2D}(E) = g_v \frac{Wk}{\pi} = g_v \frac{W}{\lambda_B(E)/2} , \qquad (2.28)$$

where $\lambda_B = 2\pi/k$ is the de Broglie wavelength of electrons at energy, E. We now see how to interpret eqn. (2.24); $M(E)$ is simply the number of electron half wavelengths that fit into the width of the conductor. This occurs because the boundary conditions insist that the wavefunction goes to zero at the two edges of the conductor.

We can now re-write eqns. (2.15) as

$$\boxed{\begin{aligned} N &= \int \frac{D(E)}{2} (f_1 + f_2) \, dE \\ I &= \frac{2q}{h} \int M(E) (f_1 - f_2) \, dE \end{aligned}} \qquad (2.29)$$

which shows that to compute the number of electrons and the current, we need two different quantities, $D(E)$ and $M(E)$. The density-of-states is a familiar quantity. For parabolic energy bands, we know that the 1D, 2D, and 3D densities-of-states are given by

$$1D: \quad D(E) = D_{1D}(E)L = \frac{L}{\pi\hbar} \sqrt{\frac{2m^*}{(E - E_c)}} H(E - E_c)$$

$$2D: \quad D(E) = D_{2D}(E)A = A \frac{m^*}{\pi\hbar^2} H(E - E_c) \qquad (2.30)$$

$$3D: \quad D(E) = D_{3D}(E)\Omega = \Omega \frac{m^* \sqrt{2m^*(E - E_c)}}{\pi^2 \hbar^3} H(E - E_c) ,$$

where L is the length of the 1D channel, A is the area of the 2D channel, Ω is the volume of the 3D channel, and H is the Heaviside step function. We now also know how to work out the corresponding results for $M(E)$; for parabolic energy bands they are

$$M(E) = M_{1D}(E) = H(E - E_c)$$

$$M(E) = W M_{2D}(E) = W g_v \frac{\sqrt{2m^*(E - E_c)}}{\pi \hbar} H(E - E_c) \qquad (2.31)$$

$$M(E) = A M_{3D}(E) = A g_v \frac{m^*}{2\pi \hbar^2}(E - E_c) H(E - E_c),$$

where W is the width of the 2D channnel and A is the cross sectional area of the 3D channel. Figure 2.3 compares the density-of-states and number of modes (conducting channels) in 1D, 2D, and 3D for the case of parabolic energy bands $(E(k) = E_c + \hbar^2 k^2/2m^*)$.

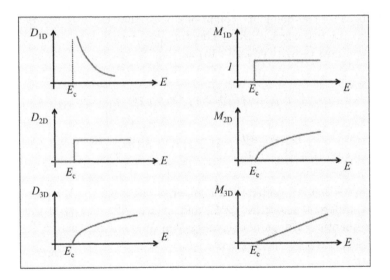

Fig. 2.3.   Comparison of the density-of-states, $D(E)$, and number of channels, $M(E)$, in 1D, 2D, and 3D. Parabolic energy bands are assumed in each case.

We can summarize the main points of this section as follows.

(1) The density-of-states vs. E is used to compute carrier densities.
(2) The number of modes (channels) vs. E is used to compute the current.

(3) The number of modes at energy, E, is proportional to the average velocity (in the direction of transport) at energy, E, times the density-of-states, $D(E)$.

(4) $M(E)$ depends on the band structure and on dimensionality.

Although we assumed parabolic energy bands to work out examples, the main results, eqn. (2.25), are general. See Lecture 10 — the graphene case study to see how to work out $M(E)$ for graphene. For general band structures, a numerical procedure can be used [5].

2.4 Transmission

Figure 2.2 showed how electrons flow from contact 1 to contact 2 under ballistic conditions. Figure 2.4 shows the diffusive case.

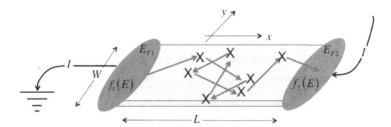

Fig. 2.4. A simple, 2D electronic device with channel width, W and length, L. In this case, diffusive transport is assumed — the channel is many mean-free-paths long.

Electrons injected from contact 1 (or 2) undergo a random walk. Some of these random walks terminate at the injecting contact and some at the other contact. If there is a positive voltage on contact 2, then a few more of the random walks terminate on contact 2. The average distance between scattering events is known as the mean-free-path. Transport is "diffusive" when the sample length is much longer than the mean-free-path. A key parameter in our model is the quantity $\gamma \pi D/2$, which we have seen is $M(E)$ for ballistic transport. The broadening, γ, is related to the transit time according to $\gamma = \hbar/\tau$. We expect the transit time to increase when transport is diffusive, so $\gamma \pi D/2$ will decrease. In this section, we will show that for diffusive transport $\gamma \pi D/2 = M(E)T(E)$, where $T(E) \leq 1$ is known as the "transmission".

For ballistic transport, there is a distribution of transit times because carriers are injected into the channel at different angles. Accordingly, we evaluated γ from the average transit time and found

$$\gamma(E) = \frac{\hbar}{\langle \tau(E) \rangle} \,, \tag{2.32}$$

where

$$\langle \tau(E) \rangle = \frac{L}{\langle v_x^+(E) \rangle} = \frac{L}{v(E) \langle \cos \theta \rangle} = \frac{L}{v(E) (2/\pi)} \,. \tag{2.33}$$

Our challenge now is to determine $\langle \tau(E) \rangle$ for the case of diffusive transport.

Consider a device with a very long channel ($L \gg \lambda$), then Fick's Law of diffusion should apply. If we inject electrons from contact 1 and collect them from contact 2, then the current in our 2D device should be given by

$$J = qD_n \frac{dn_s}{dx} \quad \text{A/cm} \,. \tag{2.34}$$

As shown in Fig. 2.5, there is a finite concentration of injected electrons at $x = 0$, $\Delta n_s(0)$, and for a long channel, $\Delta n_s(L) \to 0$. The electron profile is linear because no recombination-generation is assumed. The total number of electrons in the device is $N = n_s(0)WL/2$, where W is the width of the conductor in the direction normal to current flow, and L is the length. From our definition of transit time, we find

$$\tau = \frac{qN}{I} = \frac{Wq\Delta n_s(0)L/2}{WqD_n\Delta n_s(0)/L} = \frac{L^2}{2D_n} \,, \tag{2.35}$$

where we have used $I = JW$ and $dn_s/dx = \Delta n_s(0)/L$. We conclude that the diffusive transit time is

$$\tau_D = \frac{L^2}{2D_n} \,, \tag{2.36}$$

while the ballistic transit time was

$$\tau_B = \frac{L}{\langle v_x^+ \rangle} \,. \tag{2.37}$$

Putting this all together, we find

$$\gamma(E)\pi \frac{D(E)}{2} = \frac{\hbar}{\tau_D} \pi \frac{D}{2} = \frac{\hbar}{\tau_B} \pi \frac{D}{2} \times \frac{\tau_B}{\tau_D} \equiv M(E)T(E) \,, \tag{2.38}$$

where

$$T(E) = \frac{\tau_B}{\tau_D} \,. \tag{2.39}$$

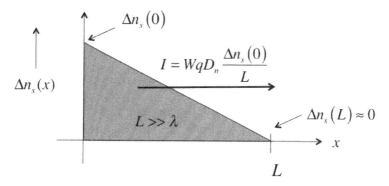

Fig. 2.5. Illustration of diffusion in a channel many mean-free-paths long.

We see that in the presence of scattering, we just need to replace $M(E)$ by $M(E)T(E)$.

To evaluate $T(E)$, we use eqn. (2.39) with eqns. (2.36) and (2.37) and find

$$T(E) = \frac{2D_n}{L \langle v_x^+ \rangle} . \tag{2.40}$$

The diffusion coefficient describes the random walk of electrons; it is related to the carrier velocity and the mean-free-path for backscattering, λ, according to

$$D_n = \frac{\langle v_x^+ \rangle \lambda}{2} \quad \mathrm{cm^2/s} . \tag{2.41}$$

(This expression is not obvious. You can check that it is dimensionally correct but will have to wait until Lecture 6 for the derivation and for a precise definition of the mean-free-path for backscattering.) Finally, using eqns. (2.40) and (2.41), we find a simple expression for the transmission:

$$T(E) = \frac{\lambda}{L} \ll 1 . \tag{2.42}$$

As expected, the product $\gamma \pi D/2 = M(E)T(E)$ is greatly reduced from its ballistic value.

Our "derivation" of $T(E)$ here is no more than a plausibility argument. As we will discuss in Lecture 6, the transmission is the probability that an electron at energy, E injected from contact 1 exits in contact 2 (or vice versa). It must be a number between 0 and 1. Equation (2.42) is accurate in the diffusive limit that we have considered ($L \gg \lambda$), but it fails when L is short. The correct, general expression is

$$\boxed{T(E) = \frac{\lambda(E)}{\lambda(E) + L}}, \qquad (2.43)$$

which reduces to eqn. (2.42) for the diffusive limit of $L \gg \lambda$, but for the ballistic limit of $L \ll \lambda$ it approaches 1. This expression is reasonable, but we will see in Lecture 6 that it can be derived with relatively few assumptions and that it is valid not only in the ballistic and diffusive limits, but in between as well.

To summarize, we can write in general

$$\gamma(E)\pi\frac{D(E)}{2} = M(E)T(E), \qquad (2.44)$$

with $M(E)$ being given by eqns. (2.25) and $T(E)$ by eqn. (2.43). People speak of three different transport regimes:

$$
\begin{array}{llll}
\text{Diffusive}: & L \gg \lambda & T = \lambda/L \ll 1 & \\
\text{Ballistic}: & L \ll \lambda & T \to 1 & (2.45) \\
\text{Quasi} - \text{ballistic}: & L \approx \lambda & T < 1. &
\end{array}
$$

Our simple transport model can be used to describe all three regions.

2.5   Near-equilibrium (linear) transport

To summarize, we have developed an expression for the current in a nanoscale device that can be expressed in two different ways:

$$I = \frac{2q}{h} \int \gamma(E)\pi\frac{D(E)}{2}\,(f_1 - f_2)\,dE$$

$$I = \frac{2q}{h} \int T(E)M(E)\,(f_1 - f_2)\,dE. \qquad (2.46)$$

There is no limitation to small applied biases yet, but if we apply a large bias, then there could be a lot of inelastic scattering that would invalidate our assumption that the current flows in independent energy channels. Since our interest is in near-equilibrium transport, we now simplify these equations for low applied bias.

The two Fermi functions in eqn. (2.46) are different when there is an applied bias. Recall that an applied bias lowers the Fermi level by $-qV$. If the applied bias is small, we can write

$$(f_1 - f_2) \approx -\frac{\partial f_0}{\partial E_F} \Delta E_F \, . \tag{2.47}$$

From the form of the equilibrium Fermi function,

$$f_0 = \frac{1}{1 + e^{(E - E_F)/k_B T_L}} \, , \tag{2.48}$$

we see that

$$\frac{\partial f_0}{\partial E_F} = -\frac{\partial f_0}{\partial E} \, . \tag{2.49}$$

Equations (2.49) and (2.47) can be used in eqn. (2.46) along with $\Delta E_F = -qV$ to obtain

$$I = \left[\frac{2q^2}{h} \int T(E) M(E) \left(-\frac{\partial f_0}{\partial E} \right) dE \right] V = GV \, . \tag{2.50}$$

The final result,

$$\boxed{G = \frac{2q^2}{h} \int T(E) M(E) \left(-\frac{\partial f_0}{\partial E} \right) dE \, ,} \tag{2.51}$$

is just the conductance in Ohm's Law, but now we have an expression that relates the conductance to the properties of the material. It is important to remember that this expression is valid in 1D, 2D, or 3D, if we use the appropropiate expression for $M(E)$.

2.6 Transport in the bulk

In this lecture, we have developed a model for the current or conductance of a device whose channel length may be short or long. When the channel is long, the contacts play no role, and the current is limited by the material properties of the channel. We can develop an expression for the current in a bulk conductor from either of the two forms of the current equations, eqn. (2.46). Let's use the first form.

Assuming near-equilibrium conditions, we can use eqn. (2.47) to write eqn. (2.46) as

$$I = \frac{2q}{h} \int \left[\gamma(E) \pi \frac{D(E)}{2} \left(-\frac{\partial f_0}{\partial E_F} \right) \Delta E_F \right] dE \, . \tag{2.52}$$

A bulk conductor is, by definition, in the diffusive limit, so

$$\gamma(E) = \frac{\hbar}{\tau(E)} = \frac{\hbar}{L^2/2D_n(E)}.$$ (2.53)

To be specific, let's assume a 2D conductor for which we can write

$$D(E) = WLD_{2D}(E).$$ (2.54)

Now using eqns. (2.53) and (2.54) in (2.52), we find

$$J_{nx} = I/W = \left[\int qD_n(E)D_{2D}(E)\left(-\frac{\partial f_0}{\partial E_F}\right)dE\right]\frac{\Delta E_F}{L}. \quad \text{A/cm}$$ (2.55)

Figure 2.6 illustrates how we think about a bulk resistor. In a conventional resistor, the potential and electrochemical potential (or quasi-Fermi level) drop linearly along the length. In our model device, the Fermi levels are only defined in the two contacts. Since the bulk resistor is assumed to be under low bias and near-equilibrium everywhere, we can conceptually place two contact separated by a length, $L \gg \lambda$, anywhere along the length of the resistor. The average electrochemical potential in the first contact, becomes E_{F1} for our "device", and the average electrochemical potential in the second contact, our E_{F2}. Because the electrochemical potential drops linearly with position, $\Delta E_F/L$ becomes dF_n/dx, and we can write (2.55) as

$$\boxed{J_{nx} = \sigma_n\frac{d(F_n/q)}{dx},}$$ (2.56)

where the conductivity is

$$\boxed{\sigma_n = \int q^2D_n(E)D_{2D}(E)\left(-\frac{\partial f_0}{\partial E}\right)dE.}$$ (2.57)

Equations (2.56) and (2.57) are standard results that are conventionally obtained from irreversible thermodynamics or by solving the Boltzmann Transport Equation [7]. We have obtained the standard expressions for bulk materials by assuming that the channel of our model device is much longer that a mean-free-path.

Real resistors can be linear even when quite large voltages are applied. How does this occur? It occurs because when the resistor is long, electrons do not drop down the total potential drop in one step. Instead, they continually gain a little energy and then dissipate it by emitting phonons. If the resistor is long and the voltage drop not too large, then the electrons

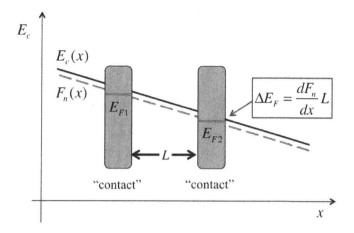

Fig. 2.6. Illustration of how a near-equilibrium bulk conductor is conceptually treated as a device with two contacts.

are always near-equilibrium, and we can conceptually divide up the resistor into sub-devices, as sketched in Fig. 2.6, where only a fraction of the potential drop occurs.

Equation (2.56) can also be written differently. Conventional semiconductor theory tells us that for a non-degenerate, n-type semiconductor

$$n_s = N_{2D}e^{(F_n - Ec)/k_B T_L}$$

$$N_{2D} = g_v \frac{m^* k_B T_L}{\pi \hbar^2}$$

$$F_n = E_c + k_B T_L \ln \frac{N_{2D}}{n_s} \tag{2.58}$$

$$\sigma_n = n_s q \mu'_n .$$

(Note that the units of the 2D conductivity, σ_n, are Siemens or $1/\Omega$.) Using eqns. (2.58), the current equation, (2.56) becomes

$$\boxed{J_{nx} = n_s q \mu_n \mathcal{E}_x + q D_n \frac{dn_s}{dx}}, \tag{2.59}$$

where

$$\boxed{\frac{D_n}{\mu_n} = \frac{k_B T_L}{q}} \tag{2.60}$$

is the Einstein relation. Equation (2.59) is the well-known "drift-diffusion" equation, which is often the starting point for analyzing semiconductor devices [4]. We see that it assumes steady-state, near-equilibrium, non-degenerate conditions (and we have also assumed a uniform temperature along the resistor).

Finally, you may be wondering: "What about holes?" In standard semiconductor physics, the conduction and valence bands are described by two different electrochemical potentials (or quasi-Fermi levels), F_n and F_p. This occurs because we have two separate populations of carriers that are in equilibrium with carriers in the same band but not with carriers in the other band. The recombination-generation processes that couple the two populations are typically slow in comparison to the scattering processes that establish equilibrium within each band. For electrons in the conduction band, we have

$$J_{nx} = \sigma_n \frac{d(F_n/q)}{dx}$$

$$\sigma_n = \int q^2 D_n(E) D_{2D}(E) \left(-\frac{\partial f_0}{\partial E_F} \right) dE \qquad (2.61)$$

$$f_0 = \frac{1}{1 + e^{(E - F_n(x))/k_B T_L}},$$

and for electrons in the valence band, we have

$$J_{px} = \sigma_p \frac{d(F_p/q)}{dx}$$

$$\sigma_p = \int q^2 D_p(E) D_{2D}(E) \left(-\frac{\partial f_0}{\partial E_F} \right) dE \qquad (2.62)$$

$$f_0 = \frac{1}{1 + e^{(E - F_p(x))/k_B T_L}}.$$

The total current is the sum of the contributions from each band. It is important to note that these equations refer to *electrons* in both the conduction and valence bands. The occupation factor, f_0, describes the probability that an *electron* state is occupied. It is often useful to visualize the resulting current flow in the valence band in terms of holes, but the expressions that we used were derived for electrons, and we did not inquire as to whether they were in the conduction or valence bands because it does not matter.

2.7   Summary

This has been a long lecture, but the final result is a simple one that we shall see is very powerful. Equation (2.51) describes the conductance of a linear resistor very generally. The conductance is proportional to some fundamental constants, $(2q^2/h)$, which we will see in the next lecture is the "quantum of conductance", that is associated with the contacts. The conductance is related to the number of conducting channels at energy, E, $M(E)$, and to the transmission, $T(E)$, which is the probability that an electron with energy, E, injected from one contact exits to the other contact. We find the total conductance by integrating the contributions of all of the energy channels. Equation (2.51) is valid in 1D, 2D, or 3D — we simply need to use the correct expressions for $M(E)$. It is valid for very short (ballistic) resistors or very long (diffusive) resistors and for the region in between. The next lecture will begin with eqn. (2.51). Finally, it should be mentioned, that we have assumed isothermal conditions — the two contacts are at the same temperature. The implications of temperature gradients will be discussed in Lectures 4 and 5.

2.8   References

The approach summarized in this chapter is described much more fully by Datta.

[1] Supriyo Datta, *Electronic Transport in Mesoscopic Systems*, Cambridge Univ. Press, Cambridge, UK, 1995.

[2] Supriyo Datta, *Quantum Transport: Atom to transistor*, Cambridge Univ. Press, Cambridge, UK, 2005.

[3] Supriyo Datta, *Lessons from Nanoelectronics: A new approach to transport theory*, World Scientific Publishing Company, Singapore, 2011.

The standard approach to semiconductor devices is clearly presented by Pierret.

[4] Robert F. Pierret *Semiconductor Device Fundamentals*, Addison-Wesley, Reading, MA, USA, 1996.

Jeong describes how to evaluate $M(E)$ for an arbitrary $E(k)$.

[5] Changwook Jeong, Raseong Kim, Mathieu Luisier, Supriyo Datta, and Mark Lundstrom, "On Landauer vs. Boltzmann and Full Band vs. Effective Mass Evaluation of Thermoelectric Transport Coefficients", *J. Appl. Phys.*, **107**, 023707, 2010.

Two classic references on low-field transport are:

[6] J.M. Ziman, *Principles of the Theory of Solids*, Cambridge Univ. Press, Cambridge, U.K., 1964.

[7] N.W. Ashcroft and N.D. Mermin, *Solid–State Physics*, Saunders College, Philadelphia, PA, 1976.

A collection of additional resources on carrier transport can be found at:

[8] M. Lundstrom, S. Datta, and M.A. Alam, "Electronics from the Bottom Up", http://nanohub.org/topics/ElectronicsFromTheBottomUp, 2011.

Hear a lecture on this chapter at:

[9] M. Lundstrom, "General Model for Transport", http://nanohub.org/topics/LessonsfromNanoscience, 2011.

Lecture 3

Resistance: Ballistic to Diffusive

Contents

3.1   Introduction

We are now ready to use the general model introduced in Lecture 2. We
will consider a simple problem, determining the resistance of 1D, 2D, and
3D resistors beginning with very short (ballistic) resistors, then treating
conventional, diffusive resistors, and finally treating the entire ballistic to
diffusive spectrum. As sketched in Fig. 3.1, in a 1D resistor (a "nanowire")
electrons are free to move in only one dimension. In a 2D resistor (in which
electrons are said to be confined in a "quantum well") electrons are free
to move in two dimensions. In a 3D resistor, electrons are free to move in
all three dimensions. According to conventional semiconductor theory (e.g.
[1]), we would write the resistances as

$$1\text{D}: \quad R_{1D} = \rho_{1D} \, L \qquad \rho_{1D} = \frac{1}{n_l q \mu_n}$$

$$2\text{D}: \quad R_{2D} = \rho_{2D} \frac{L}{W} \qquad \rho_{2D} = \frac{1}{n_s q \mu_n} \qquad (3.1)$$

$$3\text{D}: \quad R_{3D} = \rho_{3D} \frac{L}{A} \qquad \rho_{3D} = \frac{1}{n q \mu_n} \, .$$

(Note that the resistivities, ρ, have different units in different dimensions, and the carrier densities, n_l, n_s, and n are per unit length, area, and volume respectively.) These expressions are reasonable; the resistance is proportional to the length of the resistor in each case. It is inversely proportional to the width in 2D or cross-sectional area in 3D because increasing W or A is like adding resistors in parallel. We shall see however, that these equations are not always correct — even for such a simple device, interesting things can happen.

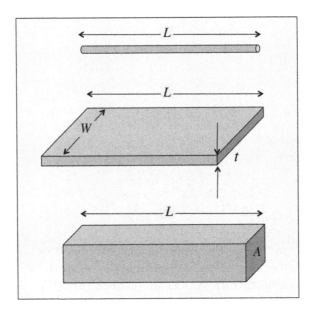

Fig. 3.1.   Sketch of 1D, 2D, and 3D resistors. In this chapter, we will focus on 2D resistors, but the same techniques apply in 1D and 3D as well.

Our starting point is the Landauer expression for the conductance, eqn. (2.51), which is repeated below

$$G = \frac{2q^2}{h} \int T(E)M(E)\left(-\frac{\partial f_0}{\partial E}\right) dE \qquad S = (1/\Omega). \qquad (3.2)$$

Equation (3.2) is valid in 1D, 2D, and 3D, as long as we use the correct expression for the number of conducting channels, $M(E)$. To be specific, consider 2D, for which it is convenient to write

$$G = \frac{1}{\rho_{2D}}\frac{W}{L} = \sigma_s \frac{W}{L}. \qquad (3.3)$$

We will see that for wide and long diffusive conductors, the sheet conductance, σ_s, is independent of W and L. For short conductors, σ_s becomes a function of L and for narrow conductors, the conductance increases with W in a stepwise manner.

In this lecture, we will focus on 2D resistors, just to make the discussion concrete, but similar considerations apply to 1D and to 3D resistors. Recall also that the $-(\partial f_0/\partial E)$ term in eqn. (3.2) came from a Taylor series expansion of $(f_1 - f_2)$ assuming that the temperature of the two contacts was the same. As we will see in the next lecture, there are two driving "forces" for current flow, differences in the Fermi levels of the two contacts (caused by different voltages) and differences in the temperatures of the two contacts. In this lecture we assume that the two contacts are at the same temperature.

3.2   2D resistors: ballistic

We begin with a ballistic resistor, for which $T(E) = 1$. The term, $M(E) = W M_{2D}(E)$ was given in eqn. (2.25) for general bands and by (2.31) for parabolic energy bands, so we just need to understand the term, $-(\partial f_0/\partial E)$, in eqn. (3.2). We refer to this term as the "Fermi window".

Figure 3.2 is a sketch of $f_0(E)$ and $-(\partial f_0/\partial E)$ vs. E. We see that $(-\partial f_0/\partial E)$ is significant only for an energy range of a few $k_B T_L$ near the Fermi level. It is readily shown that the area under the $-(\partial f_0/\partial E)$ vs. E curve is one, so for low temperatures, we may write

$$-\frac{\partial f_0}{\partial E} \approx \delta\left(E - E_F\right). \tag{3.4}$$

Finally, using eqn. (3.4) with $T(E) = 1$ in eqn. (3.2), we find

$$\boxed{G^{\text{ball}} = \frac{2q^2}{h} M\left(E_F\right),} \tag{3.5}$$

which is a general expression valid in any dimension. If the number of channels is small, then we can simply count them, and we find that the conductance or resistance cannot be any value, it is quantized according to

$$R^{\text{ball}} = \frac{1}{M(E_F)} \frac{h}{2q^2} = \frac{12.9 \text{ k}\Omega}{M(E_F)}. \tag{3.6}$$

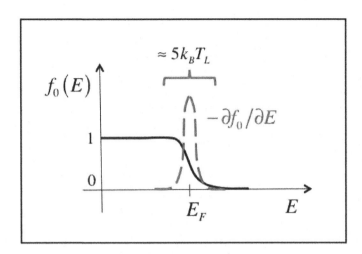

Fig. 3.2.   Sketch of the Fermi function and its derivative vs. energy. The function, $-(\partial f_0/\partial E)$, is called the "Fermi window" for conduction.

Note that the ballistic resistance is independent of length, as expected for ballistic transport.

The fact that resistance is quantized is well-established experimentally. See, for example, Fig. 3.3, which shows experimental results. The resistor is a 2D electron gas formed at the interface of AlGaAs and GaAs. The width of the resistor is controlled electrostatically by reverse-biased Schottky junctions. The mobility of the electrons is very high (because the electrons reside in an undoped GaAs layer and because the temperature is low), so ballistic transport is expected. As the width was electrically varied, the measured conductance was seen to increase in discrete steps according to eqn. (3.5). Quantized conductance has been observed in many different systems. The experiment shown in Fig. 3.3 was done at low temperature to achieve near ballistic transport, but modern devices are so short that these effects are becoming important at room temperature in some systems.

Wide, 2D ballistic resistors: $T_L = 0$ K

When W is many electron half-wavelengths, then the number of channels is large, and it is no longer easy to count them. In this limit, we have for

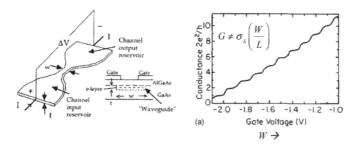

Fig. 3.3. Experiments of van Wees, *et al.* experimentally demonstrating that conductance is quantized. Left: sketch of the device structure. Right: measured conductance. (Data from: B. J. van Wees, *et al.*, *Phys. Rev. Lett.* **60**, 848851, 1988. Figures from D. F. Holcomb, "Quantum electrical transport in samples of limited dimensions", *Am. J. Phys.*, **67**, pp. 278-297, 1999. Reprinted with permission from *Am. J. Phys.* Copyright 1999, American Association of Physics Teachers.)

parabolic energy bands from eqn. (2.31)

$$M(E_F) = W M_{2D}(E_F) = W g_v \frac{\sqrt{2m^*(E_F - E_c)}}{\pi \hbar} . \qquad (3.7)$$

It is convenient to relate M_{2D} to the sheet carrier density, n_s because n_s is often known in experiments. In momentum space, all states with $k < k_F$ are occupied at $T_L = 0$. The 2D area of k-space occupied is πk_F^2, each state takes up an area in k-space of $(2\pi)^2/A$, and there is a spin degeneracy of 2, so

$$n_s = g_v \frac{\pi k_F^2}{(2\pi)^2} \times 2 = g_v \frac{k_F^2}{2\pi} , \qquad (3.8)$$

where g_v is the valley degeneracy. By solving this equation for k_F and using eqn. (2.28), we find

$$M_{2D}(E_F) = \sqrt{\frac{2n_s}{g_v \pi}} . \qquad (3.9)$$

Equation (3.9) relates the number of channels at the Fermi energy to the sheet carrier density. It is interesting to note that this result does not assume a particular band structure — only that the band is isotropic. To relate k_F to E_F, however, we have to assume a band structure. For example, for parabolic energy bands, the wavevector, k_F is found by solving

$$\frac{\hbar^2 k_F^2}{2m^*} = (E_F - E_c) . \qquad (3.10)$$

Wide, 2D ballistic resistors: $T_L > 0$ K

At low temperatures, the approximation, eqn. (3.4), works well, but near room temperature and above, we must usually work out the integral in eqn. (3.2). Using the definition of the Fermi function, eqn. (2.1), we find

$$G_{2D}^{\text{ball}} = \frac{2q^2}{h} \int W M_{2D}(E) \left(-\frac{\partial}{\partial E} \right) \frac{1}{1 + e^{(E-E_F)/k_B T_L}} dE . \qquad (3.11)$$

Integrals of this type appear frequently in semiconductor physics, so let's work this one out as an example.

From the form of the Fermi function, we see that

$$\left(-\frac{\partial}{\partial E} \right) = \left(+\frac{\partial}{\partial E_F} \right) , \qquad (3.12)$$

which allows us to move the derivative outside the integral in eqn. (3.11) to obtain

$$G_{2D}^{\text{ball}} = \frac{2q^2}{h} \frac{W g_v \sqrt{2m^*}}{\pi \hbar} \left(\frac{\partial}{\partial E_F} \right) \int_0^\infty \frac{\sqrt{(E - E_c)}}{1 + e^{(E-E_F)/k_B T_L}} dE , \qquad (3.13)$$

where we have used eqn. (3.7) for $M_{2D}(E)$. Next, we change variables by defining

$$\begin{aligned} \eta &\equiv (E - E_c)/k_B T_L \\ \eta_F &\equiv (E_F - E_c)/k_B T_L , \end{aligned} \qquad (3.14)$$

and find

$$G_{2D}^{\text{ball}} = \frac{2q^2}{h} \frac{W g_v \sqrt{2m^* k_B T_L}}{\pi \hbar} \left(\frac{\partial}{\partial \eta_F} \right) \int_0^\infty \frac{\sqrt{\eta}}{1 + e^{\eta - \eta_F}} d\eta . \qquad (3.15)$$

The integral in eqn. (3.15) cannot be done analytically, but integrals of this type occur so often in semiconductor work that they have been given a name — Fermi-Dirac integrals. In this case, the integral in eqn. (3.15) is proportional to

$$\mathcal{F}_{1/2}(\eta_F) \equiv \frac{2}{\sqrt{\pi}} \int_0^\infty \frac{\eta^{1/2}}{1 + e^{\eta - \eta_F}} d\eta , \qquad (3.16)$$

which is the Fermi-Dirac integral of order one-half. Finally, one property of a Fermi-Dirac integral is that when differentiated with respect to its argument, the order of a Fermi-Dirac integral is reduced by one. Putting this all together, we find

$$G_{2D}^{\text{ball}} = \frac{2q^2}{h} \frac{W g_v \sqrt{2m^* k_B T_L}}{\pi \hbar} \left(\frac{\sqrt{\pi}}{2} \right) \mathcal{F}_{-1/2}(\eta_F) = \frac{2q^2}{h} \langle W M_{2D} \rangle , \quad (3.17)$$

where

$$\langle M \rangle = \langle WM_{2D} \rangle = \left(\frac{\sqrt{\pi}}{2} \right) WM_{2D}(k_B T_L)\mathcal{F}_{-1/2}(\eta_F), \qquad (3.18)$$

where $WM_{2D}(k_B T_L)$ is $WM_{2D}(E - E_c)$ evaluated at an energy of $E - E_c = k_B T_L$. Comparing eqns. (3.17) and (3.5), we see that the conductance at finite temperatures has the same form as the $T_L = 0$ K result — we just replace $M(E_F)$ by $\langle M \rangle$ as defined in eqn. (3.18). We interpret $\langle M \rangle$ as the number of channels in the Fermi window, $(-\partial f_0 / \partial E)$.

When analyzing experiments, it is often easier to determine n_s than E_F, but the two are related, so given n_s, we can find E_F. For parabolic bands, the relation is

$$n_s = \int_0^\infty D_{2D}(E)f_0(E)dE = g_v \frac{m^* k_B T_L}{\pi \hbar^2}\mathcal{F}_0(\eta_F) = N_{2D}\mathcal{F}_0(\eta_F). \quad (3.19)$$

We have worked out one example assuming a 2D resistor with parabolic energy bands. Similar integrals with Fermi-Dirac integrals of various orders appear when working out problems in other dimensions and for other (i.e. nonparabolic) dispersions. Familiarity with a few properties of Fermi-Dirac integrals is helpful when working out such integrals.

Fermi-Dirac integrals

The Fermi-Dirac integral of order j is defined as

$$\mathcal{F}_j(\eta_F) \equiv \frac{1}{\Gamma(j+1)} \int_0^\infty \frac{\eta^j d\eta}{1 + e^{\eta - \eta_F}}, \qquad (3.20)$$

where the Γ-function is defined for integer arguments of zero or greater as

$$\Gamma(n) = (n-1)!. \qquad (3.21)$$

We also have the following useful relations,

$$\begin{aligned} \Gamma(1/2) &= \sqrt{\pi} \\ \Gamma(p+1) &= p\Gamma(p) \end{aligned} \qquad (3.22)$$

For non-degenerate semiconductors, the Fermi level is several $k_B T_L$ below the band edge, so $\eta_F = (E_F - E_c)/k_B T_L \ll 0$. Under these conditions, Fermi-Dirac integrals of any order reduce to exponentials:

$$\mathcal{F}_j(\eta_F) \to e^{\eta_F} \qquad \eta_F \ll 0. \qquad (3.23)$$

Another useful property that we have already seen involves taking the derivative of a Fermi-Dirac integral,

$$\frac{d\mathcal{F}_j(\eta_F)}{d\eta_F} = \mathcal{F}_{j-1}(\eta_F). \qquad (3.24)$$

These few definitions and rules are all we need for most semiconductor problems. One warning — don't confuse the "script" Fermi-Dirac integral as defined in eqn. (3.20) with the "Roman" Fermi-Dirac integral, $F_j(\eta_F)$, which does not include the Γ-function normalization. For a good introduction to Fermi-Dirac Integrals — including approximations and scripts to compute them, see [2].

Exercise 3.1: Analysis of a silicon MOSFET

Let's see how we can use the results of this section to analyze the performance of modern field-effect transistors. Figure 3.4 shows the measured I–V characteristics of a 60 nm channel length silicon MOSFET. Let's focus on the low drain voltage (near-equilibrium) region and the $V_G = 1.2$ V characteristic (the top line). For this condition, the measured carrier density and channel resistance (after subtracting out the parasitic series resistances of the source and drain contacts) are

$$n_s \approx 6.7 \times 10^{12} \text{ cm}^{-2}$$

$$R_{ch} \approx 215 \ \Omega\text{-}\mu\text{m}$$

$$\mu_{\text{eff}} \approx 260 \text{ cm}^2/\text{V-s}.$$

The two questions we ask are: 1) How many conduction channels carry the current? and 2) How close is the channel resistance to the ballistic limit? Before we can answer these questions, we need to understand a little bit about the Si band structure. Recall that in bulk Si, there are six equivalent conduction band minima [1], but quantum confinement lifts the degeneracy so that the lowest subband is two-fold degenerate ($g_v = 2$) with an effective mass of $m^* = m_t = 0.19m_0$ [3]. We can do the calculation in three different ways. First, we can assume $T_L = 0$ K, which makes the math easy, but the $T_L = 0$ K assumption is not so good at room temperature. Second, we can assume Maxwell-Boltzmann (non-degenerate) carrier statistics, which also results in simple math, but above the threshold voltage, the non-degenerate assumption is not so good. The third way to do the calculation is without either assumption, which entails evaluating Fermi-Dirac integrals. In this

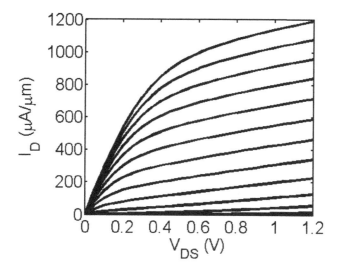

Fig. 3.4. Measured $I_D - V_{DS}$ characteristic of an n-channel silicon MOSFET. (Data from: Changwook Jeong, Dimitri A. Antoniadis and Mark Lundstrom, "On Back-Scattering and Mobility in Nanoscale Silicon MOSFETs", *IEEE Trans. Electron Dev.*, **56**, 2762-2769, 2009.)

simple example, we shall consider just the first approach, which should give us a good feel for the numbers.

Assuming $T_L = 0$ K, eqn. (3.9), gives $M_{2D}(E_F) \approx 150/\mu$m. Consider a minimum size transistor with $W/L = 2$. Since $L = 60$ nm, we find $M_{2D}(E_F) \approx 18$. A rather small number of channels carry the current in a minimum sized MOSFET. To find the corresponding ballistic resistance of this MOSFET, we use eqn. (3.6) to find $R_{2D}^{\text{ball}} \approx 90$ Ω-μm, which is almost one-half of the channel resistance of this MOSFET.

3.3   2D resistors: diffusive to ballistic

In the previous section, we assumed ballistic transport, $T(E) = 1$. In this section, we will first assume diffusive transport, $T(E) = \lambda(E)/L$. Our general approach works for 1D, 2D, and 3D resistors, but as in the previous section, we'll just work out the 2D case here. Our starting point is eqn. (3.2) in 2D and in the diffusive limit.

$$G_{2D}^{\text{diff}} = \left(\frac{2q^2}{h} \int \lambda(E) M_{2D}(E) \left(-\frac{\partial f_0}{\partial E} \right) dE \right) \frac{W}{L} \quad (1/\text{Ohm}). \quad (3.25)$$

The ratio, W/L, is consistent with conventional transport theory (e.g. eqn. (3.1)). In our Landauer picture, the W enters because the number of channels is proportional to W, and the L comes because in the diffusive limit, the transmission is proportional to $1/L$. Since the integral in eqn. (3.25) is trivial for $T_L = 0$ K, we begin there.

Wide, 2D diffusive resistors: $T_L = 0$ K

Using eqn. (3.4), we find that for $T_L = 0$ K, eqn. (3.25) becomes

$$G_{2D}^{\text{diff}} = \frac{2q^2}{h} M_{2D}(E_F) \lambda(E_F) \frac{W}{L} = \frac{\lambda(E_F)}{L} G_{2D}^{\text{ball}}, \qquad (3.26)$$

so if we know the mean-free-path, it is easy to compute the diffusive conductance.

We have discussed the ballistic and diffusive conductances, but it is also easy to cover the entire ballistic to diffusive spectrum. By using $T(E) = \lambda(E)/(\lambda(E) + L)$, we find

$$G_{2D} = \frac{2q^2}{h} W M_{2D}(E_F) \frac{\lambda(E_F)}{\lambda(E_F) + L} = \frac{\lambda(E_F)}{\lambda(E_F) + L} G_{2D}^{\text{ball}}, \qquad (3.27)$$

or in terms of resistance,

$$R = \left(1 + \frac{L}{\lambda(E_F)}\right) R^{\text{ball}}. \qquad (3.28)$$

Equation (3.28) shows that $R \propto L$ in the diffusive limit and R is independent of L in the ballistic limit.

Wide, 2D diffusive resistors: $T_L > 0$ K

At finite temperatures, we should work out the integral in eqn. (3.25). By multiplying and dividing eqn. (3.25) by

$$\langle M_{2D} \rangle \equiv \int M_{2D}(E) \left(-\frac{\partial f_0}{\partial E}\right) dE,$$

we can re-arrange it as,

$$G_{2D}^{\text{diff}} = \frac{2q^2}{h} \langle M_{2D} \rangle \langle\langle \lambda \rangle\rangle \frac{W}{L} = \frac{\langle\langle \lambda \rangle\rangle}{L} G_{\text{ball}}^{2D}, \qquad (3.29)$$

where $\langle M \rangle$ was defined in eqn. (3.18), and the average mean-free-path is defined as

$$\langle\langle \lambda \rangle\rangle \equiv \frac{\int \lambda(E) M_{2D}(E) \left(-\frac{\partial f_0}{\partial E}\right) dE}{\int M_{2D}(E) \left(-\frac{\partial f_0}{\partial E}\right) dE} = \frac{\langle M\lambda \rangle}{\langle M \rangle} . \tag{3.30}$$

Equation (3.29) has the same form as (3.26). For $T_L = 0$ K, we just replace $\langle M \rangle$ with $M(E_F)$ and $\langle\langle \lambda \rangle\rangle$ with $\lambda(E_F)$. The single and double brackets are just to remind us that we are dealing with two different, specially defined averages.

To actually evaluate the average mean-free-path, we need to assume a band structure and specify $\lambda(E)$. A simple way to write $\lambda(E)$ for some common scattering mechanisms is in so-called power-law form [4],

$$\lambda(E) = \lambda_0 \left(\frac{E - E_c}{k_B T_L}\right)^r , \tag{3.31}$$

where r is a characteristic exponent describing a particular scattering process and λ_0 is a (typically temperature-dependent) constant. For example, in 3D, $r = 0$ for carrier scattering by acoustic phonons, and $r = 2$ for ionized impurity scattering.

With only a little straight-forward (and not too tedious) math, we can work out the integral in eqn. (3.30) to find

$$\langle\langle \lambda \rangle\rangle = \lambda_0 \times \left(\frac{\Gamma\left(r + 3/2\right)}{\Gamma\left(3/2\right)}\right) \times \left(\frac{\mathcal{F}_{r-1/2}(\eta_F)}{\mathcal{F}_{-1/2}(\eta_F)}\right) . \tag{3.32}$$

If $r = 0$ the mean-free-path is independent of energy and $\langle\langle \lambda \rangle\rangle = \lambda_0$.

Finally, we should consider the entire ballistic to diffusive spectrum. For an energy dependent mean-free-path, the result is a bit complicated, but for a constant mean-free-path, the result is much like eqn. (3.28)

$$R = R^{\text{ball}} \left(1 + \frac{L}{\lambda_0}\right) . \tag{3.33}$$

Exercise 3.2: Analysis of a silicon MOSFET (continued)

In the previous example, we estimated the ballistic channel resistance by assuming $T_L = 0$ K (not because it is a good assumption, just to keep the math simple) and found $R^{\text{ball}} \approx 90 \ \Omega\text{-}\mu\text{m}$. The measured channel resistance is 215 $\Omega\text{-}\mu\text{m}$, so from eqn. (3.33) we have

$$215 = \left(1 + \frac{L}{\lambda_0}\right) 90 \to \lambda_0 \approx 40 \, \text{nm}$$

Our $T_L = 0$ K assumption for analyzing room temperature data results in a mean-free-path that is a little too large (assuming Maxwell Boltzmann statistics gives 15 nm), but this simple calculation does illustrate the point that modern silicon transistors are in the quasi-ballistic transport regime — neither fully ballistic nor fully diffusive. A more careful analysis would involve Fermi-Dirac integrals and also consider the possibility that multiple subbands are occupied.

3.4   Discussion

In this lecture we discussed the evaluation of the resistance of a 2D conductor. The techniques and concepts apply to other dimensions as well. In this section, we will discuss a few topics relating to resistors and resistivity.

A few words about mobility

The conventional description of resistance begins with eqns. (3.1), but these equations do not apply to ballistic or quasi-ballistic resistors, and it is not always clear about how to evaluate the mobility. Our approach to transport begins with eqn. (3.2), which applies from the ballistic to diffusive limits. There has been no need for us to discuss mobility. Although mobility is a commonly-used concept, it can often be confusing. For example, eqns. (3.1) tell us that the conductivity is proportional to the electron density times the mobility, but the term, $-\left(\partial f_0/\partial E\right)$, in eqn. (3.2) (the Fermi window) ensures that only electrons near the Fermi level contribute to current flow. For an n-type semiconductor, this is sometimes all of the carriers in the conduction band (non-degenerate semiconductors), but sometimes only a small fraction of them (degenerate semiconductors). Nevertheless, because the mobility concept is so frequently used, we should discuss it.

The best way to define the mobility is to begin with eqn. (3.2) and equate it to the conventional expression,

$$G_{2D} = \frac{2q^2}{h} \int T(E)M(E) \left(-\frac{\partial f_0}{\partial E}\right) dE \equiv n_s q \mu \frac{W}{L}, \qquad (3.34)$$

from which we find the mobility as

$$\mu_{\text{app}} \equiv \frac{1}{n_s} \frac{2q}{h} \int T(E) L \, M_{2D}(E) \left(-\frac{\partial f_0}{\partial E} \right) dE , \qquad (3.35)$$

which we take as the definition of mobility, not the simple Drude expression,

$$\mu = \frac{q\tau}{m^*} , \qquad (3.36)$$

where τ is the momentum relaxation time. We label the mobility in eqn. (3.35) an "apparent" mobility because eqn. (3.35) is defined for ballistic transport as well as diffusive. For example, setting $T(E) = 1$, we find the "ballistic mobility" as

$$\mu_{\text{ball}} = \frac{1}{n_s} \frac{2q}{h} \int L M_{2D}(E) \left(-\frac{\partial f_0}{\partial E} \right) dE . \qquad (3.37)$$

Similarly, setting $T(E) = \lambda(E)/L$, we find the traditional, diffusive mobility as

$$\boxed{\mu_{\text{diff}} = \frac{1}{n_s} \frac{2q}{h} \int \lambda(E) M_{2D}(E) \left(-\frac{\partial f_0}{\partial E} \right) dE .} \qquad (3.38)$$

The concept of a ballistic mobility was introduced by Shur [5] and can be useful in device analysis. It allows us to use traditional expressions in the ballistic limit simply by replacing the actual mobility with the ballistic mobility. Comparing eqn. (3.37) with (3.38), we see that the ballistic mobility is the diffusive mobility with the mean-free-path replaced by the length of the resistor. Recall that in the contacts, strong scattering maintains equilibrium. An electron injected into a ballistic channel last scattered in the first contact, and then scatters next in the second contact. The distance between scattering events is, therefore, the channel length, so it makes physical sense that it plays the role of the mean-free-path in the ballistic mobility.

Increasingly in nanoscale electronics, problems lie between the ballistic and diffusive limits. In this case, $T(E) = \lambda(E)/(\lambda(E) + L)$, and we can show that the apparent mobility is

$$\frac{1}{\mu_{\text{app}}} = \frac{1}{\mu_{\text{diff}}} + \frac{1}{\mu_{\text{ball}}} , \qquad (3.39)$$

which looks like a traditional Mathiessen's Rule for combining mobilities [4]. Another way to look at this is to use eqn. (3.38) for the diffusive mobility but replace the mean-free-path with an apparent mean-free-path,

$$\frac{1}{\lambda_{\text{app}}} = \frac{1}{\lambda} + \frac{1}{L} \, . \tag{3.40}$$

The apparent mean-free-path is the actual mean-free-path due to scattering or the length of the channel — whichever is shorter.

Assuming parabolic energy bands, the mobilities for general conditions can be worked out in terms of Fermi-Dirac integrals. We leave that as an exercise. Consider the simpler case of $T_L = 0$ K. The electron density in the conduction band is

$$n_s = g_v \frac{m^*}{\pi \hbar^2} \left(E_F - E_c \right) = D_{2D} \left(E_F - E_c \right) , \tag{3.41}$$

and we can also show

$$M_{2D} = \frac{h}{4} \left\langle v_x^+ \right\rangle D_{2D} , \tag{3.42}$$

where

$$\left\langle v_x^+ \right\rangle = \frac{2}{\pi} v_F , \tag{3.43}$$

with v_F being the Fermi velocity.

Using eqns. (3.37), (3.38), (3.41), and (3.42), we find

$$\begin{aligned} \mu_{\text{diff}} &= \frac{D_{\text{diff}}}{\left(E_F - E_c \right)/q} \\ \mu_{\text{ball}} &= \frac{D_{\text{ball}}}{\left(E_F - E_c \right)/q} , \end{aligned} \tag{3.44}$$

where the diffusion coefficients are given by

$$\begin{aligned} D_{\text{diff}} &= \frac{\left\langle v_x^+ \right\rangle \lambda(E_F)}{2} \\ D_{\text{ball}} &= \frac{\left\langle v_x^+ \right\rangle L}{2} \, . \end{aligned} \tag{3.45}$$

Equations (3.44) look like Einstein relations between the mobility and diffusion coefficient with $(E_F - E_c)/q$ playing the role of $k_B T_L/q$ since we have assumed $T_L = 0$ K. Note that in eqn. (3.45), D_{diff} is what is generally called the diffusion coefficient, and we have also defined a "ballistic diffusion coefficient"!

Exercise 3.3: Analysis of a silicon MOSFET (continued)

Let's return to our MOSFET example and estimate the ballistic mobility of this transistor. The measured mobility comes from a long channel MOSFET, so it is the traditional, diffusive mobility. Let's estimate the ballistic mobility. We'll assume $T_L = 0$ K again, to keep the math simple. Under this assumption, the ballistic mobility, eqn. (3.44) can be expressed in terms of the inversion layer density, n_s according to

$$\mu_{\text{ball}} = \frac{2q}{h} L \sqrt{2g_v/\pi n_s} . \tag{3.46}$$

Inserting numbers, we find, $\mu_{\text{ball}} \approx 1200$ cm^2/V-s, which is larger than the diffusive mobility. According to eqn. (3.39), the apparent mobility of this MOSFET will be a little lower than the bulk mobilty because of the ballistic mobility. (The precise numbers here should be taken with caution. We are assuming $T_L = 0$ K to keep the math simple but the mobility was measured at 300 K.)

Ways to write the conductivity

According to eqn. (3.26), the 2D diffusive conductivity (also called the sheet conductance) at $T_L = 0$ K is

$$\sigma_s = \frac{2q^2}{h} M_{2D} (E_F) \lambda (E_F) . \tag{3.47}$$

You will see this expression written in several different ways, so it's worth mentioning some of the common forms.

In Lecture 2, we saw that

$$M_{2D}(E) = \frac{h}{4} \langle v_x^! \rangle D_{2D}(E)$$
$$\langle v_x^+ \rangle = \frac{2}{\pi} v , \tag{3.48}$$

and in Lecture 6 we will learn that

$$\lambda(E) = \frac{\pi}{2} v(E) \tau_m(E) , \tag{3.49}$$

where τ_m is the momentum relaxation time, the time between electron scattering events. Using these expressions, we can rewrite eqn. (3.47) as

$$\sigma_s = q^2 D_{2D} (E_F) \frac{v^2(E_F) \tau(E_F)}{2} , \tag{3.50}$$

which is a form that you will commonly see. By defining the electron diffusion coefficient as

$$D_n = \frac{v^2(E_F)\tau(E_F)}{2} \,,$$ (3.51)

we can rewrite eqn. (3.50) as

$$\sigma_s = q^2 D_{2D}(E_F) D_n(E_F) \,,$$ (3.52)

which is also a common way to write the sheet conductance.

Finally, let's discuss one more way. Assuming parabolic energy bands, we have

$$\frac{1}{2} m^* v(E_F)^2 = (E_F - E_c) \,.$$ (3.53)

Using this and eqn. (3.41) for n_s, we can write eqn. (3.50) as

$$\sigma_s = n_s q \mu_n \,,$$ (3.54)

where the mobility is

$$\mu_n = \frac{q\tau(E_F)}{m^*} \,.$$ (3.55)

Equation (3.55) is a familiar result, but not a good starting point for analysis in general. For example, in research, we frequently encounter problems for which the parabolic band assumption that leads to an effective mass, m^* is not appropriate.

We summarize this discussion by collecting the various ways to write the 2D conductivity:

$$\boxed{\begin{aligned} \sigma_s &= \frac{2q^2}{h} M_{2D}(E_F)\, \lambda(E_F) \\[2mm] \sigma_s &= q^2 D_{2D}(E_F)\, \frac{v^2(E_F)\tau(E_F)}{2} \\[2mm] \sigma_s &= q^2 D_{2D}(E_F)\, D_n(E_F) \\[2mm] \sigma_s &= n_s q \mu_n = n_s q \left(\frac{q\tau(E_F)}{m^*} \right) \,. \end{aligned}}$$ (3.56)

While these expressions are all equivalent, they provide different perspectives. According to the second and third forms, what's important is the density of states and velocity or diffusion coefficient at the Fermi level. According to the last expression, the conductivity is proportional to the total carrier density. Of course the term, $-(\partial f_0/\partial E)$, in the conductivity expression tells us that the current is carried by electrons near the Fermi level (i.e. inside the Fermi window), but the position of the Fermi level can be related to the total carrier density. It is useful to understand these different perspectives because sometimes a problem that is difficult from one perspective is easy when viewed from a different perspective.

Finally, we should extend our discussion to finite temperatures where eqn. (3.47) becomes

$$\sigma_s = \frac{2q^2}{h} \int M_{2D}\left(E\right) \lambda\left(E\right) \left(-\frac{\partial f_0}{\partial E}\right) dE. \tag{3.57}$$

It is convenient to define a "differential conductivity" so that this equation can be written as

$$\boxed{\begin{aligned} \sigma_s &= \int \sigma_s'\left(E\right) dE \\ \sigma_s' &= \frac{2q^2}{h} M_{2D}\left(E\right) \lambda\left(E\right) \left(-\frac{\partial f_0}{\partial E}\right). \end{aligned}} \tag{3.58}$$

To find the total conductivity, we add up the contributions from each energy channel. You will find eqn. (3.58) written in several different ways, as in eqn. (3.56).

Where is the power dissipated?

For any resistor, the power dissipated is $P_d = VI = V^2/R$. The power input to the electron system from the battery is typically dissipated by electron-phonon scattering, which transfers the energy to the lattice and heats it up. For a ballistic resistor, there is no scattering in the channel, but the power dissipated is still V^2/R. Where is this power dissipated? Not surprisingly, if the answer isn't in the channel, then it must be in the contacts.

We know that current flows in the Fermi window, the energy range where $f_1 - f_2$ is non zero. At $T_L = 0$ K, the current flows between the two Fermi levels, as illustrated in Fig. 3.5. For finite temperatures and small applied biases, the current flows in the region where $-(\partial f_0/\partial E)$ is significant. As shown in Fig. 3.5, when an electron leaves contact 1, it leaves a hole (an empty state) in the contact. These electrons enter contact 2 with some excess energy (we say that they are "hot" electrons), which they lose by

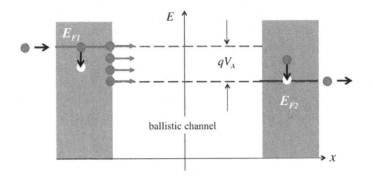

Fig. 3.5.   Illustration of how power is dissipated in a ballistic resistor.

inelastic scattering in contact 2. The average energy of the electrons that enter contact 2 is about $qV_A/2$, so about half of the power is dissipated in contact 2. Current flows near the Fermi level, so charge neutrality in contact 2 is maintained when an electron at the Fermi energy leaves contact 2 and travels around the external circuit until it enters contact 1. The electron enters contact 1 at the Fermi energy, and inelastic scattering processes fill up the hole that was created when the first electron was injected into the channel. On average, the energy of the hole is about $qV_A/2$, so the other half of the power is dissipated in contact 1.

We conclude that in a ballistic channel near-equilibrium, one-half of the power is dissipated in each of the two contacts.

Where does the voltage drop?

In a diffusive resistor, the voltage drops linearly across the length of the resistor. Where does the voltage drop in a ballistic resistor? Not surprisingly, the answer is at the contacts. For a thorough discussion of this topic, see Chapter 2 in Datta [6]. Here, we just explain why it is reasonable to expect the voltage to drop across the two contacts.

Recall that a voltmeter measures differences in the Fermi levels (electrochemical potentials) of the two contacts. For example, in a p-n junction, there is an electrostatic potential difference between the p and n-type regions in equilibrium, but a voltmeter reads zero because the Fermi level is the same in the two contacts.

Figure 3.6 is a sketch of an energy band diagram for a ballistic resistor under bias. $E_{F2} = E_{F1} - qV_A$, so a voltmeter across the two contacts

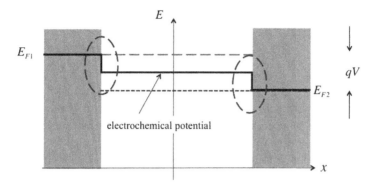

Fig. 3.6. Energy band diagram of a ballistic resistor under bias illustrating how we associate the internal voltage drop with the change in the electrochemical potential (also known as the quasi-Fermi level).

will register a voltage of V_A, but where does this voltage drop within the resistor?

Inside contact 1, there is one, well-defined Fermi level, E_{F1}, and inside contact 2, there is a second, well-defined Fermi level, E_{F2}. Inside the device however, there are two Fermi levels. Some states are filled by the source. Since they are in equilibrium with the source, they are filled according to a Fermi function with the source Fermi level. The other set of states is filled by the drain according to a Fermi function with the drain Fermi level. If we compute the average Fermi level, the electrochemical potential, it has the shape shown by the solid line in Fig. 3.6. We see that one-half of the electrochemical potential drop occurs at the first contact and the other half at the second contact. Since a voltmeter responds to changes in the electrochemical potential, we conclude that the voltage drops equally at the two contacts. For this reason, the ballistic resistance, 12.9 k$\Omega/\langle M \rangle$, is often called the "quantum contact resistance".

Resistors in 1D and 3D

Our general expression for conductance, eqn. (3.2), works for the 1D, 2D, or 3D resistors sketched in Fig. 3.1, as long as we properly count the modes (channels) for current flow. Consider again the 2D resistor sketched in Fig. 3.1. It is long in the direction of current flow, but finite in the width and thickness directions. Whenever electrons are confined in a potential well, their energies are quantized into "particle in a box" states according to

$$\epsilon_n = \frac{\hbar^2 \pi^2 n^2}{2m^* a^2},$$

(3.59)

where a is the width of the potential well, and n is an integer. Each of these "subbands" is occupied according to its location with respect to the Fermi level and is a channel for current flow. If the thickness of the 2D sheet is very thin, then the corresponding subbands are widely spaced in energy, and we can count them. (We have been implicitly assuming that only the lowest subband is occupied.) If the width, W, of the 2D sheet is large, then the corresponding subbands are closely spaced in energy, and many are occupied. In this case, the number of subbands is proportional to W. To include all subbands, we write

$$M(E) = W M_{2D}(E) = \sum_{n=1}^{N} W g_v \frac{\sqrt{2m^*(E - \epsilon_n)}}{\pi \hbar},$$

(3.60)

where the sum is over the subband in the vertical confinement direction.

Now consider the 1D resistor in Fig. 3.1; it is like a very narrow 2D resistor. If the width and thickness are both small, then all of the subbands are widely spaced in energy, and we can simple count them,

$$M(E) = M_{1D}(E) = \text{No. of subbands at energy, } E.$$

(3.61)

Finally, if both the width and thickness of the resistor are large, all of the subbands are closely spaced in energy, and

$$M(E) = A M_{3D}(E) = g_v \frac{m^*(E - E_c)}{2\pi \hbar^2}.$$

(3.62)

For a 1D (nanowire) resistor, we have strong quantum confinement in two dimensions, and $M(E)$ is given by eqn. (3.61). For a 2D (planar) resistor, we have strong quantum confinement in one dimension, and $M(E)$ is given by eqn. (3.60). And for a 3D resistor, there is no quantum confinement and $M(E)$ is given by eqn. (3.62). (Note that eqns. (3.60) and (3.62) assume parabolic energy bands.) Having defined $M(E)$, we can evaluate eqn. (3.2) in any dimension.

It is convenient to rewrite eqn. (3.2) as

$$G = \frac{2q^2}{h} \langle\langle T \rangle\rangle \langle M \rangle$$

$$\langle M \rangle = \int M(E) \left(-\frac{\partial f_0}{\partial E} \right) dE \qquad (3.63)$$

$$\langle\langle T \rangle\rangle = \frac{\int T(E) M(E) \left(-\frac{\partial f_0}{\partial E} \right) dE}{\int M(E) \left(-\frac{\partial f_0}{\partial E} \right) dE} .$$

If we assume a constant mean-free-path (just to keep things simple) we can write the resistance as

$$G = \frac{2q^2}{h} \frac{\lambda_0}{\lambda_0 + L} \langle M \rangle . \qquad (3.64)$$

Assuming parabolic energy bands, we can write in 1D:

$$\langle M_{1D} \rangle = \sum_i \mathcal{F}_{-1} (\eta_{Fi}) , \qquad (3.65)$$

where

$$\eta_{Fi} = \frac{E_F - \epsilon_i}{k_B T_L} , \qquad (3.66)$$

and the index, i, refers to the various subbands. At $T_L = 0$ K, the 1D expressions reduce to

$$G_{1D} = \frac{2q^2}{h} \frac{\lambda_0}{\lambda_0 + L} \times \text{No. of subbands at energy, } E_F . \qquad (3.67)$$

For Maxwell-Boltzmann statistics, we find

$$G_{1D} = n_l q \mu_{\text{app}} \frac{1}{L} , \qquad (3.68)$$

where

$$\mu_{\text{app}} = \frac{D_n}{(k_B T_L / q)}$$

$$D_n = v_T \lambda_{\text{app}} / 2$$

$$v_T = \sqrt{(2 k_B T_L) / \pi m^*} \qquad (3.69)$$

$$\frac{1}{\lambda_{\text{app}}} = \frac{1}{\lambda_0} + \frac{1}{L} .$$

In 2D, we find

$$\langle M \rangle = W \langle M_{2D} \rangle = \frac{\sqrt{\pi}}{2} W M_{2D} \left(k_B T_L \right) \sum_i \mathcal{F}_{-1/2} \left(\eta_{Fi} \right) , \qquad (3.70)$$

where

$$M_{2D} \left(k_B T_L \right) = g_v \frac{\sqrt{2m^* k_B T_L}}{\pi \hbar} . \qquad (3.71)$$

At $T_L = 0$ K, eqn. (3.65) simplifies to

$$G_{2D} = \frac{2q^2}{h} \frac{\lambda_0}{\lambda_0 + L} W M_{2D} \left(E_F \right) . \qquad (3.72)$$

For Maxwell-Boltzmann statistics, we find

$$G_{2D} = n_s q \mu_{\text{app}} \frac{W}{L} . \qquad (3.73)$$

Finally, in 3D, the expressions become:

$$\langle M \rangle = A \langle M_{3D} \rangle = A M_{3D} \left(k_B T_L \right) \mathcal{F}_0 \left(\eta_F \right) , \qquad (3.74)$$

where

$$M_{3D} \left(k_B T_L \right) = g_v \frac{m^* k_B T_L}{2\pi \hbar^2} \qquad (3.75)$$

and

$$\eta_F = \frac{E_F - E_c}{k_B T_L} . \qquad (3.76)$$

At $T_L = 0$ K, the 3D expressions reduce to

$$G_{3D} = \frac{2q^2}{h} \frac{\lambda_0}{\lambda_0 + L} A M_{3D} \left(E_F \right) . \qquad (3.77)$$

For Maxwell-Boltzmann statistics, we find

$$G_{3D} = n q \mu_{\text{app}} \frac{A}{L} . \qquad (3.78)$$

Exercise 3.4: A 1D example

Let's end this chapter with a simple example. Carbon nanotubes are almost ideal 1D conductors. Figure 3.7 is the measured IV characteristic of a metallic carbon nanotube. For low voltages, the current is linear, and we can read off the conductance as 22 μS. For this material, the $T_L = 0$ K approximation is good even at moderate temperatures, so we can use eqn. (3.67). There is a valley degeneracy of two for the carbon nanotube, and assuming that one subband is occupied, we find the ballistic resistance to be $G_B = 154$ μS. From eqn. (3.67), we can estimate the mean-free-path to be $\lambda_0 = 167$ nm, which is much less than the 1 μm length of the carbon nanotube, so transport in this carbon nanotube is diffusive.

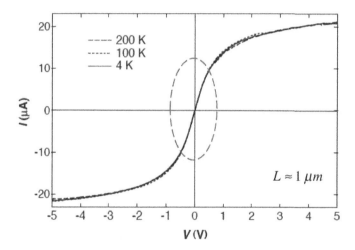

Fig. 3.7. Measured I–V characteristic of a metallic carbon nanotube. Essentially identical measurements at $T_L = 4, 100$, and 200 K are shown. (Zhen Yao, Charles L. Kane, and Cees Dekker, "High-Field Electrical Transport in Single-Wall Carbon Nanotubes", *Phys. Rev. Lett.*, **84**, 2941-2944, 2000. Reprinted with permission from *Phys. Rev. Lett.* Copyright 2000, American Physical Society.)

3.5 Summary

Our goal in this lecture has been to learn how to use eqn. (3.2), which applies quite generally when the temperature is uniform. We learned that:

(1) Conductors display a finite resistance, even in the absence of scattering in the resistor. The ballistic resistance sets a lower limit to resistance no matter how short the resistor is. This ballistic limit is becoming important in practical, room temperature devices.

(2) The ballistic (quantum contact) resistance is quantized, and the quantum of resistance is $h/2q^2$.

(3) Transport from the ballistic to diffusive limit is easily treated by using the transmission.

(4) Resistors in 1D, 2D, or 3D can all be treated with a common formalism.

For the most part, we assumed parabolic energy bands with an effective mass, m^*, when working out results. It's important to realize, however, that the assumption of parabolic energy bands is not necessary. For an example of working with a much different bandstructure, see Lecture 10, where we develop similar expressions for graphene, a material with linear, not parabolic energy bands.

Finally, the key point to remember is that when you encounter a new material or nanostructure and need to know its resistance, the place to begin is at eqn. (3.2) not at eqn. (3.1).

3.6   References

For an introduction to semiconductor theory as used for analyzing electronic devices, see:

[1] Robert F. Pierret *Advanced Semiconductor Fundamentals, 2nd Ed.*, Vol. VI, Modular Series on Solid-State Devices, Prentice Hall, Upper Saddle River, N.J., USA, 2003.

The essentials of Fermi-Dirac integrals are described in:

[2] R. Kim and M.S. Lundstrom, "Notes on Fermi-Dirac Integrals", 3rd Ed., https://www.nanohub.org/resources/5475.

For a discussion of quantum confinement in silicon MOSFETs, how the valley degeneracy is lifted, etc., see:

[3] M. Lundstrom, "ECE 612: Nanoscale Transistors (Fall 2008). Lecture

4: Polysilicon Gates/QM Effects", http://nanohub.org/resources/5364, 2008.

Mathiessen's Rule for combining mobilities is discussed in Chapter 2 of Lundstrom, and power law scattering is disucssed on pp. 67 and 137.

[4] Mark Lundstrom, *Fundamentals of Carrier Transport 2nd Ed.*, Cambridge Univ. Press, Cambridge, UK, 2000.

The concept of ballistic mobility was introduced by Michael Shur in

[5] M.S. Shur, Low Ballistic Mobility in GaAs HEMTs, *IEEE Electron Dev. Lett.*, **23**, 511-513, 2002.

For an excellent discussion of power dissipation and voltage drops in ballistic conductors, see:

[6] Supriyo Datta, *Electronic Transport in Mesoscopic Systems*, Cambridge Univ. Press, Cambridge, UK, 1995.

Lecture 4

Thermoelectric Effects: Physical Approach

Contents

4.1 Introduction

Thermoelectric (TE) devices convert heat into electricity or electric power into cooling (or heating) power. In this lecture, our focus is on the physics of thermoelectricity; only heuristic mathematical arguments are used. In Lecture 5, we will present a more formal, mathematical derivation.

In Lecture 2 we saw that the electrical current in the bulk is

$$J_{nx} = \sigma_n \frac{d\left(F_n/q\right)}{dx} \qquad (\text{A/m}^2),\tag{4.1}$$

which can also be written as

$$\frac{d\left(F_n/q\right)}{dx} = \rho_n J_{nx}.\tag{4.2}$$

Recall that when the carrier density is uniform, $d(F_n/q)/\,dx = \mathcal{E}_x$, where $\mathcal{E}_x$ is the electric field. In this lecture, we will assume bulk, diffusive transport in 3D, but the same considerations apply to 1D and 2D as well. Similar equations also apply in the ballistic and quasi-ballistic regimes, as will be discussed in Lecture 5. The question we are concerned with in this lecture is:

How do these equations change in the presence of a temperature gradient? We will see that the answer can be written in two ways:

$$J_{nx} = \sigma_n \frac{d\left(F_n/q\right)}{dx} - S_n\sigma_n \frac{dT_L}{dx}$$

$$\frac{d\left(F_n/q\right)}{dx} = \rho_n J_{nx} + S_n \frac{dT_L}{dx},$$

(4.3)

where S_n is the *Seebeck coefficient* (also called the *thermopower*) in V/K.

Thermoelectricity involves the flow of charge and heat, so in addition to the equation for the charge current, we need an equation for the heat current too. Since heat flows down a temperature gradient, we expect an equation of the form,

$$J_{Qx} = -\kappa \frac{dT_L}{dx} \qquad \text{W/m}^2.$$

(4.4)

How does eqn. (4.4) change in the presence of an electric current? The answer can also be written in two ways:

$$J_{Qx} = T_L S_n \sigma_n \frac{d\left(F_n/q\right)}{dx} - \kappa_0 \frac{dT_L}{dx}$$

$$J_{Qx} = \pi_n J_{nx} - \kappa_n \frac{dT_L}{dx}.$$

(4.5)

In eqns. (4.5)

$$\pi_n = T_L S_n$$

(4.6)

is the *Peltier cofficient* and

$$\kappa_n = \kappa_0 - S_n^2 \sigma_n T_L$$

(4.7)

is the *electronic thermal conductivity* for zero current flow. Similarly, κ_0 is the electronic thermal conductivity when there is no change in the quasi-Fermi level (short circuit conditions). It is important to understand that both electrons and the lattice (phonons) carry heat. These equations refer *only* to the portion of the heat carried by the electrons.

Our goal in this lecture is to understand the physical origin of the Seebeck and Peltier coefficients and how they are related to the properties of the semiconductor. The approach used here is similar to that of Datta [1].

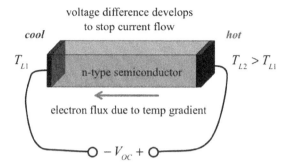

Fig. 4.1. Sketch of an n-type semiconductor slab with its two contacts open-circuited and with contact 2 hotter than contact 1.

4.2   Electric current flow: Seebeck effect

Let's discuss the Seebeck effect first. To be specific, consider the n-type semiconductor sketched in Fig. 4.1.

When a temperature difference exists, we expect that the electrons should diffuse from the hot end to the cold end, but a positive voltage must develop to stop the flow. The open-circuit voltage is the Seebeck voltage. Similarly, if the semiconductor were p-type, a negative open-circuit voltage would be measured.

We can also think about the Seebeck effects in terms of Fermi levels. Fig. 4.2 shows a sketch of the Fermi function at two different temperatures. The width of the transition region is a few $k_B T_L$; the higher the temperature, the wider the transition region. Recall from Lecture 2 that the current depends on $(f_1 - f_2)$. If we are dealing with a lightly-doped, n-type semiconductor, then the states carrying the current are above the Fermi level where $f_2 > f_1$ in Fig. 4.2. To stop the current, a positive voltage must develop on contact 2 to lower its Fermi level and make $f_1 = f_2$. Alternatively, if we are dealing with a lightly doped, p-type semiconductor, then the states carrying the current lie below the Fermi level where $f_1 > f_2$. To stop the current, a negative voltage must develop on contact 2 to raise its Fermi level and make $f_1 = f_2$.

As this discussion shows, the origin of the Seebeck voltage is easy to appreciate. The sign of the Seebeck voltage is positive (hot side voltage – cold side voltage) for n-type conduction and negative for p-type conduction. This fact can be used to determine the type of a semiconductor. Next, we turn to the question of understanding what controls the magnitude of the Seebeck voltage.

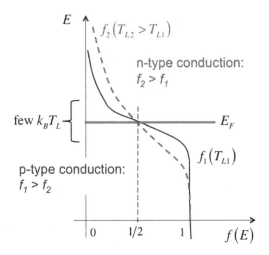

Fig. 4.2. Sketch of equilibrium Fermi functions vs. energy for two different temperatures.

Figure 4.3(a) is a sketch of an energy band diagram for an n-type semi-conductor in equilibrium where $E_{F1} = E_{F2}$ and $T_{L1} = T_{L2}$. For any state, (E, x), in the device, the probability that it is occupied from contact one, f_1, is the same as the probability that it is occupied from contact 2, f_2, so $f_1 = f_2$ and no current flows. (Strictly speaking, the notion that states inside the device are occupied according to the Fermi levels of the two contacts only applies to very short devices, but we can view the contacts as "conceptual contacts" inside a very long device, as was illustrated in Fig. 2.6.)

Figure 4.3(b) shows the resistor with a positive bias on contact 2. Assume first that $T_{L2} = T_{L1}$. Now for any state in the device, (E, x), $f_1 > f_2$, so electrons flow from left to right producing a current in the negative x direction. Alternatively, we could have seen this by noting that the positive voltage on contact 2 attracts electrons, causing them to flow from left to right and producing the current in the negative x direction.

Finally, consider Fig. 4.3(b) when the voltage and temperature of contact 2 are both larger than the corresponding values on contact 1. The fact that contact 2 is hotter than contact 1 should cause electrons to flow from contact 2 to contact 1, but the fact that the voltage at contact 2 is higher than contact 1 causes electrons to flow from contact 1 to contact 2. Under open circuit conditions, these two effects cancel, and the current is zero.

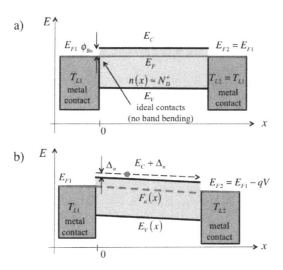

Fig. 4.3. Energy band diagrams for an n-type semiconductor for two different conditions: (a) equilibrium, and (b) $V_2 > V_1$. In the second case, T_{L2} may be the same as or different than T_{L1}.

Next, consider a short section of the resistor sketched in Fig. 4.1. Assume that contact 2 is hotter than contact 1 and that the voltage on contact 2 is such that the current is zero (the energy band diagram is like Fig. 4.3(b)). Now consider an energy state located at position $x = 0$ and at the average energy at which the current flows, $E_c(0) + \Delta_n$. (The precise value of Δ_n depends on band structure and scattering physics, but for non-degenerate conditions, it is typically $\approx 2k_B T_L$.) No current flows if this state has an equal probability of being occupied by the right and left contacts (i.e. $f_1 = f_2$). The first probability is just the Fermi function of contact 1. For a short section of the resistor, the second probability is just the Fermi function of the second contact. (Here again we assume that contact 2 is a fictitious, internal contact as discussed in Chapter 2 Sec. 6.) Open circuit occurs when

$$f_1 [E_c(0) + \Delta_n] = f_2 [E_c(0) + \Delta_n] . \tag{4.8}$$

Making use of the definition of the Fermi function, eqn. (4.8) becomes

$$\frac{1}{1 + e^{(E_c + \Delta_n - E_{F1})/k_B T_{L1}}} = \frac{1}{1 + e^{(E_c + \Delta_n - E_{F1} + q\delta V)/k_B T_{L2}}} , \tag{4.9}$$

which can be solved to find

$$\delta V = -S_n \delta T_L , \tag{4.10}$$

where $\delta T_L = T_{L1} - T_{L2}$ and

$$S_n = -\frac{E_c(0) + \Delta_n - E_{F1}}{qT_{L1}} = \frac{E_J - E_{F1}}{(-q)T_{L1}}. \tag{4.11}$$

In eqn. (4.11), $E_J = E_c(0) + \Delta_n$ is the average energy at which the current flows. The Seebeck coefficient is simply related to the difference between the average energy at which current flows and the Fermi level. Equations (4.10) and (4.11) apply to each section, dx of the resistor and simply add to the resistive voltage drop due to current flow another contribution due to the temperature difference, so eqn. (4.3) becomes

$$\boxed{\begin{aligned} \frac{d(F_n/q)}{dx} &= \rho_n J_{nx} + S_n \frac{dT_L}{dx} \\ S_n(T_L) &= \left(\frac{k_B}{-q}\right)\left(\frac{E_c - E_F}{k_B T_L} + \delta_n\right), \end{aligned}} \tag{4.12}$$

where $\delta_n = \Delta_n/k_B T_L$. Under open-circuit conditions, eqn. (4.12) can be solved for

$$\Delta V = -\int_{T_{L1}}^{T_{L2}} S_n(T_L)dT_L, \tag{4.13}$$

which should be compared to eqn. (4.10) for the differential element. While we still don't know how to compute δ_n, we do now understand where eqn. (4.3) comes from.

Exercise 4.1: Seebeck coefficient of Ge

Figure 4.4 is a plot of the measured Seebeck coefficient for Ge at room temperature as a function of the separation between the Fermi level and the band edge. For this exercise:

(1) Show that the measured results are consistent with eqn. (4.12) and that they imply that δ_n is small and constant for a non-degenerate semiconductor.
(2) Show that δ_n increases when E_F moves above the bottom of the conduction band.
(3) Discuss how to write eqns. (4.12) for a p-type semiconductor.

According to eqn. (4.12), $|S_n|$ should drop linearly as E_F increases, as long as δ_n is constant. This appears to describe the results for $E_F \ll E_c$. According to eqn. (4.12), $|S_n| = 86\mu V/K \times (-\eta_F + \delta_n)$, where $\eta_F =$

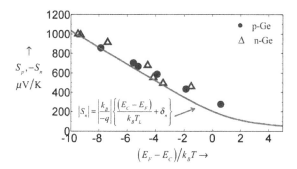

Fig. 4.4. Measured Seebeck coefficient for n- and p-type Ge at $T_L = 300$ K. The line is a calculation using methods in Lecture 5 assuming parabolic energy bands and a constant mean-free-path for backscattering. (Data taken from T.H. Geballe and G.W. Hull, "Seebeck Effect in Germanium", *Physical Review*, **94**, 1134, 1954.)

$(E_F - E_c)/k_B T_L$. At $\eta_F = -10$, Fig. 4.4 shows that $|S_n| \approx 1000$ μV/K, which implies that $\delta_n \approx 2$. We will see in Lecture 5 that this is a reasonable value for a non-degenerate semiconductor.

If δ_n were always constant, then S_n should change sign when $E_F > E_c + 2k_B T_L$. The fact that this does not happen tells us that δ_n must increase as E_F moves into the conduction band. At $\eta_F = 4$, Fig. 4.4 shows that $|S_n| \approx 50$ μV/K. Using eqn. (4.12), we find that $\delta_n \approx 4.6$, which is more than twice its value under non-degenerate conditions. As was illustrated in Fig. 4.3, $\delta_n = (E_J - E_c)/k_B T_L$ tells us how far from the band edge the average current flows. For a strongly degenerate semiconductor, $E_J \rightarrow E_F$, so $\delta_n \rightarrow \eta_F$ and $S_n \rightarrow 0$ for $E_F \gg E_c$.

Finally, let's consider eqn. (4.12) for a p-type semiconductor. The magnitude of the Seebeck coefficient is large when the Fermi level is far below (above) the band edge, and the sign is negative (positive) for electrons (holes). The form of the current equation in eqn. (4.3) does not change, but Seebeck coefficient in eqn. (4.12) is written as:

$$
\begin{aligned}
\frac{d\left(F_p/q\right)}{dx} &= \rho_p J_{px} + S_p \frac{dT_L}{dx} \\
S_p(T_L) &= \left(\frac{k_B}{+q}\right)\left(\frac{E_F - E_v}{k_B T_L} + \delta_p\right),
\end{aligned}
\tag{4.14}
$$

where $\delta_p = (E_v - E_J)/k_B T_L$ is a positive number that tells us how far below the valence band edge the average current flows. In general, both electrons and holes contribute to the total Seebeck coefficient. This becomes important at high temperatures or for low bandgaps. We will discuss how to include both contributions to S in Lecture 5.

Finally, we mention a subtle point that will be discussed in Lecture 5. A careful look at eqns. (4.12) and (4.14) shows that they are identical — except for the subscripts "n" or "p". The same expression describes the Seebeck coefficient due to electrons in either the conduction or valence band. It is sometimes useful to think in terms of holes and to rewrite the expression as in eqn. (4.14) to make the change in sign explicit, but to compute S, we need only one expression.

4.3   Heat current flow: Peltier effect

Figure 4.5 illustrates Peltier cooling and heating. The sample is isothermal with an electric current forced in contact 2. Electrons flow with an average (small) "drift velocity" from left to right. Since electrons scatter from phonons (lattice vibrations) in the resistor, they acquire a random (thermal) velocity (much larger than the small drift velocity). The thermal velocity is a measure of the electron temperature (which under near-equilibrium conditions is the same as the temperature of the lattice, T_L). As electrons flow from left to right, they carry their random kinetic energy (heat) with them. We see, therefore, that an electron current is accompanied by a heat current.

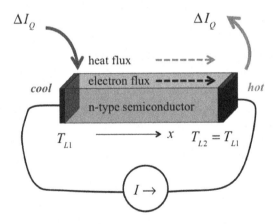

Fig. 4.5.   Illustration of the Peltier effect. The existence of an electric current causes heat to be absorbed at one contact and emitted at the other. If the direction of the current is reversed, then the contact that absorbs heat and the one that emits heat are interchanged.

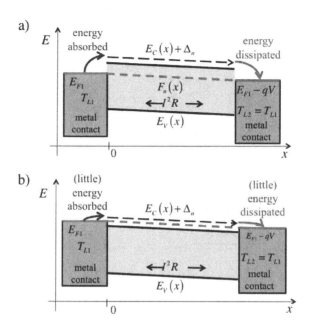

Fig. 4.6.   Illustration of heat absorption and emission in the presence of a current flow — the Peltier effect. (a) a lightly doped semiconductor, and (b) a heavily doped semiconductor.

To evaluate the heat current, consider the energy band diagram in Fig. 4.6(a). The metal contacts are strongly degenerate, so the $-(\partial f_0/\partial E)$ term in the conductance is nearly a δ-function at the Fermi energy. In the metal contacts, the current flows very close to the Fermi level. In the lightly doped semiconductor, the current flows a little above the bottom of the conduction band, at $E = E_c + \Delta_n$. The average energy at which current flows increases across the metal-semiconductor junction. Where does the energy come from? The answer is that electrons absorb thermal energy from the lattice, which is why heat is absorbed at contact 1. Just the opposite occurs at contact 2 where heat is emitted. Note that heat is also generated throughout the resistor because of the I^2R Joule heating. The Peltier heating is proportional to I and the Joule heating to I^2.

Figure 4.6(b) shows the case for a heavily doped semiconductor. In a degenerate semiconductor, the current also flows near the Fermi level. In this case, there is little or no change in the average energy at which current flows, so there is little Peltier cooling or heating at the contacts of a heavily doped semiconductor.

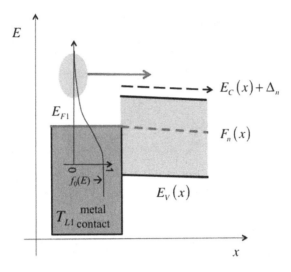

Fig. 4.7.   Illustration of heat absorption at contact 1. Electrons in the metal with high enough energy escape into the semiconductor. To replace the electrons lost, new electrons flow in (near the Fermi energy) from the contact. To replace the lost energy of the electron gas, electrons absorb thermal energy from the lattice to restore the equilibrium Fermi-Dirc distribution.

Another way to think about Peltier cooling is illustrated in Fig. 4.7, which shows the region near contact 1. Electrons with energy greater than the barrier height, ϕ_{Bn}, are thermionically emitted from the metal into the semiconductor. This process depletes the high energy electrons in the Fermi distribution, producing a non-equilibrium distribution of electrons in energy. The missing high energy electrons are replaced by electron-phonon scattering. Energy is absorbed from the lattice to move lower energy electrons up to replace the electrons that were thermionically emitted into the semiconductor. This process is much like the evaporation of a liquid, with the electrons in the metal contact playing the role of the liquid.

It is now easy to calculate the heat current. The electron flux in the $+x$-direction is $J_{nx}/(-q)$. (Since $J_{nx} < 0$ in this case, the heat flux is in the $+x$-direction). As electrons move from contact 1 to the semiconductor, each one must, on average, absorb a thermal energy of $Q = E_c(0) + \Delta_n - E_{F1}$. Accordingly, the heat flux is

$$J_{Q1} = [E_c(0) + \Delta_n - E_{F1}] \times J_{nx}/(-q) = \pi_n J_{nx}, \qquad (4.15)$$

where the *Peltier coefficient* is

$$\pi_n = \frac{[E_c(0) + \Delta_n - E_{F1}]}{-q}.$$ (4.16)

Note that $\pi_n < 0$ for the n-type semiconductor being considered here. By comparing eqn. (4.15) with (4.11), we see that

$$\pi_n = T_{L1} S_n;$$ (4.17)

there is an intimate connection between the Seebeck and Peltier coefficients. This relation, eqn. (4.17) is known as the *Kelvin relation*. Similar expressions apply at contact 2, with T_{L1} replaced by T_{L2}.

We have discussed the first term in eqn. (4.5), but not the electronic thermal conductivity, κ_n. An expression for κ_n will be derived in Lecture 5. It seems reasonable, however to expect that the electrical conductivity, σ_n, and the thermal conductivity due to electrons, κ_n, will be related because electrons carry the electrical current and the heat current. Indeed, from the equations to be developed in Lecture 5, we will see that

$$\frac{\kappa_n}{\sigma_n} = L T_L,$$ (4.18)

where L is the *Lorenz number*. Equation (4.18) is known as the *Wiedemann-Franz Law* [2]. It is not as fundamental as the Kelvin relation because it depends on details of the band structure and scattering.

As noted by Mahan and Bartkowiak [G.D. Mahan and M. Bartkowiak, "Wiedemann-Franz law at boundaries", *Appl. Phys. Lett.*, **74**, 953 (1999)], the Wiedemann-Franz "Law" is a "rule of thumb". Since electrons carry both charge and heat, we should expect a relation between σ_n and κ_n. For a typical semiconductor with parabolic energy bands and a constant mean-free-path, we find

$$L \approx 2 \times (k_B/q)^2 \qquad (\text{non} - \text{degenerate})$$
$$L \approx \frac{\pi^2}{3} \times (k_B/q)^2 \qquad (\text{degenerate}).$$ (4.19)

The term "Wiedemann-Franz Law" commonly refers to eqn. (4.18) with the specific values of Lorenz numbers as given by eqns. (4.19). But we can regard eqn. (4.18) as a generalized Wiedemann-Franz Law with a value of L that depends on details of the bandstructure and scattering physics. Lower dimension conductors, for example, can have values of L that are much different from those in eqns. (4.19). A general expression for L will be given in Lecture 5, eqn. (5.53).

Finally, we see that the thermoelectric transport parameters in eqn. (4.5) are

$$
\begin{aligned}
J_{Qx} &= \pi_n J_{nx} - \kappa_n \frac{dT_L}{dx} \\
\pi_n &= T_L S_n \\
\kappa_n &= T_L \sigma_n L \,.
\end{aligned}
\tag{4.20}
$$

We have an expression for $\pi_n = T_L S_n$ but will have to wait for Lecture 5 to develop an expression for κ_n.

4.4 Coupled flows

We have seen in the previous two sections that the basic equations of thermoelectricity are

$$
\begin{aligned}
\frac{d(F_n/q)}{dx} &= \rho_n J_{nx} + S_n \frac{dT_L}{dx} \\
J_{Qx} &= \pi_n J_{nx} - \kappa_n \frac{dT_L}{dx} \,,
\end{aligned}
\tag{4.21}
$$

where the four thermoelectric transport parameters are: 1) the electrical resistivity, $\rho_n = 1/\sigma_n$, 2) the Seebeck coefficient, S_n; 3) the Peltier coefficient, π_n; 4) the electronic heat conductivity, κ_n.

We discussed the electrical conductivity in Lecture 3. For a 3D, diffusive sample, it is given by

$$
\begin{aligned}
\sigma_n &= \int \sigma_n'(E)\, dE = \frac{2q^2}{h} \langle M_{3D} \rangle \langle\langle \lambda \rangle\rangle \\
\sigma_n'(E) &= \frac{2q^2}{h} M_{3D}(E) \lambda(E) \left(-\frac{\partial f_0}{\partial E} \right) \,.
\end{aligned}
\tag{4.22}
$$

Recall that $M_{3D}(E)$ is the number of channels for conduction at energy, E, per unit area of the resistor cross-section. The energy-dependent mean-free-path for backscattering is $\lambda(E)$, and the units of the differential conductivity are $1/(\Omega\text{-m-J})$.

We saw in Sec. 4.2 that the Seebeck coefficient is given by

$$
S_n(T_L) = \left(\frac{k_B}{-q} \right) \left(\frac{E_c - E_F}{k_B T_L} + \delta_n \right) \,,
\tag{4.23}
$$

where the parameter, $\delta_n = (E_J - E_c)/k_B T_L$, is the average energy at which current flows with respect to the bottom of the conduction band in units of $k_B T_L$. Since $\sigma'_n(E)$ tells how the current is distributed in energy, we see that

$$\delta_n = \frac{1}{k_B T_L} \left(\frac{\int (E - E_c)\, \sigma'_n(E)\, dE}{\int \sigma'_n(E)\, dE} \right). \tag{4.24}$$

We also saw in Sec. 4.3 that the Peltier coefficient is simply related to the Seebeck coefficient by the Kelvin relation,

$$\pi_n(T_L) = T_L S_n(T_L). \tag{4.25}$$

Equations (4.21) are specific examples of so-called "coupled flows". In this case, we see that a temperature gradient produces an electrical current, and an electrical current produces a flow of heat. The cross-coupling terms, S_n and π_n are fundamentally related. The Kelvin relation is a specific example of the *Onsager relations*, which relate the coupling terms in coupled flow equations [3].

Finally, we write the electronic heat conductivity as

$$\kappa_n = T_L \sigma_n L. \tag{4.26}$$

An expression for the Lorenz number, L, will be given in Lecture 5, but for parabolic bands, it is useful to remember that $L \approx 2(k_B/q)^2$ under non-degenerate conditions and $L \approx (\pi^2/3)(k_B/q)^2$ for strongly degenerate conditions. We should also keep in mind that heat is transported by lattice vibrations (phonons) and that our equations only describe the part of the heat flow due to electrons. In semiconductors, phonons carry most of the heat while in metals it is the electrons that carry most of the heat.

Exercise 4.2: Thermoelectric coefficients of Ge

To get a feel for the magnitude of the various parameters in eqns. (4.21), let's evaluate them for lightly doped, n-type Ge at room temperature. We are given the following information:

$$N_D = 10^{15}\,\text{cm}^{-3}$$
$$T_L = 300\,\text{K}$$
$$\mu_n = 3200\,\text{cm}^2/\text{V-s}.$$

Recall that the dopants in Ge are shallow and fully ionized at room temperature in lightly-doped Ge, so the equilibrium carrier density is $n_0 \approx N_D = 10^{15}$ [4]. Recall also that for a non-degenerate semiconductor, there is a simple relation between the carrier density and the Fermi level [4]

$$n_0 = N_c e^{(E_F - E_c)/k_B T_L}, \tag{4.27}$$

where the "effective density of states" for Ge is $N_c = 1.09 \times 10^{19}\,\mathrm{cm}^{-3}$ at room temperature [4]. Now let's evaluate the four transport parameters.

The first parameter is the resistivity, which is one over the conductivity. Equation (4.22) tells us how to evaluate the conductivity, but since we are given the mobility, we should write the conductivity in the alternative form, $\sigma_n = n_0 q \mu_n$. It is then an easy matter to find $\rho_n \approx 2\ \Omega$-cm.

The next parameter is the Seebeck coefficient, which according to eqn. (4.23) depends on the location of the Fermi level. We are given the carrier density, so from eqn. (4.27) we find $(E_c - E_F)/k_B T_L = \ln(N_c/n_0) \approx 9.3$. Assuming that $\delta_n = 2$, we find $S_n \approx -970\,\mu\mathrm{V/K}$.

The third parameter is the Peltier coeffient, which is simply obtained from the Kelvin relation, eqn. (4.25) as $\pi_n = -0.3\ \mathrm{W/A}$.

Finally, to determine the electronic thermal conductivity, we use eqn. (4.26). Since we are dealing with a lightly doped, 3D semiconductor with nearly parabolic energy bands, we can assume $L \approx 2(k_B/q)^2$ and find $\kappa_n \approx 2.2 \times 10^{-4}\ \mathrm{W/m\text{-}K}$.

The four thermoelectric transport parameters for this example are:

$$\rho_n \approx 2\ \Omega\text{-cm}$$
$$S_n \approx -970\ \mu\mathrm{V/K}$$
$$\pi_n \approx -0.3\ \mathrm{W/A}$$
$$\kappa_n \approx 2.2 \times 10^{-4}\ \mathrm{W/m\text{-}K}.$$

This example was rather simple, but it's worth thinking about what would change if: 1) the temperature were lowered to 77 K or 2) the temperature remains 300 K, but the doping is increased to $10^{20}\ \mathrm{cm}^3$. We should also keep in mind that the lattice conducts heat as well. The lattice thermal conductivity for Ge at 300 K is 58 W/m-K — five orders of magnitude larger than the electronic component that we have calculated here. In heavily doped semiconductors with low lattice thermal conductivity, however, the electronic component of the thermal conductivity can be a substantial fraction of the total.

4.5 Thermoelectric devices

The Seebeck effect was discovered in 1821 by the German physicist, Thomas Johann Seebeck and the Peltier effect by the French physicist, Jean Charles Athanase Peltier in 1834. During the 1950's and 60's, efficient thermoelectric materials were discovered and devices developed (notably at the Ioffe Institute in Russia [5]) — resulting in efficiencies suitable for several applications. For the next 30 years, progress was slow, but thermoelectric technology was developed for several, special purpose applications, such as power generation for deep space missions, precision temperature control of electronic devices, and picnic coolers for beverages. Researchers are currently exploring several ideas that make use of nanotechnology to enhance performance [6]. How successful this research will be remains to be seen, but progress is occurring, and each increase in performance expands the market for thermoelectric technology. A thorough discussion of thermoelectric technology would require another volume of lecture notes, but the basic concepts are easy to appreciate and will be discussed in this section.

Shown in Fig. 4.8 is a sketch of a simple thermoelectric cooler. Note that the n-type and p-type legs are connected in series electrically with the two ends shorted at the top and fed by a current source at the bottom. The operation is easy to understand in terms of the flow of electron and holes. The current flows up the n-type leg and down the p-type leg, so electrons and holes both flow from top to bottom. As the carriers flow away from the top plate, they carry heat with them. The carriers absorb heat at the two metal-semiconductor junctions, and the top plate cools. If the direction of the current is reversed, the hot and the cold plates are interchanged. Some key questions are: 1) What determines the maximum temperature difference that can be produced? 2) How much heat can be pumped from the top plate? and 3) What determined the efficiency (or COP — coefficient of performance) of this Peltier cooler?

Shown in Fig. 4.9 is a sketch of a simple thermoelectric power generator. Note that it is the same as the cooling device sketched in Fig. 4.8. In this case, however, we apply a heat source to the top plate and maintain a cooler temperature at the bottom plates. Holes flow away from the heat source, down the p-type leg, out the lead and through an external load (the light bulb in this illustration), and the current flows back in the lead at the left and up the n-type leg. The current flowing up the n-type leg completes the circuit and represents electrons flowing down the n-type leg away from the heat source. For the power generator, the key question is:

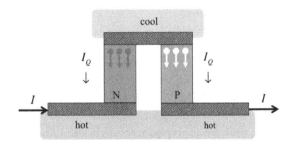

Fig. 4.8.   Schematic illustration of how a thermoelectric cooler operates.

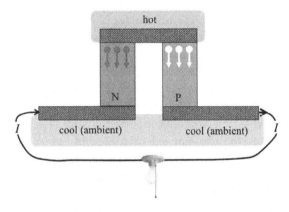

Fig. 4.9.   Schematic illustration of how a thermeoelectric power generator operates. Using this device, heat is converted into electrical power.

What determines the efficiency with which the heat flux is converted into electrical power? Before we do a simple analysis of device operation, we note that the two legs of the device are connected electrically in series but thermally in parallel. Practical devices consist of many of these n- and p-type pairs, or thermoelectric couples. The series electrical connection increases the voltage, which makes it easier to drive a large current. The fact that they are thermally in parallel increases the heat pumped or power generated.

Figure 4.10 is a sketch of a simple, one-leg TE device that we will use as an illustration of how to relate TE parameters to device performance. The device operates as a Peltier cooler. To analyze this device, we can set up a balance equation for the heat flux, Q_c extracted from the cool side.

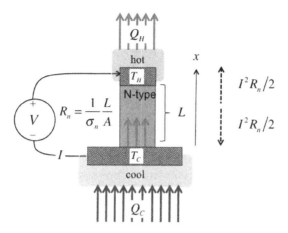

Fig. 4.10.   One leg TE device used for model calculations for a Peltier cooler.

In words, Q_c is the heat flux pumped by the Peltier effect minus the heat flux that back diffuses from the hot side and minus the Joule heating, $I^2 R$. We assume that one-half of the Joule heat flows down to the cold junction and one-half up. (The Joule heating term was not included in the basic equations of thermoelectricity, eqns. (4.21), because they describe near-equilibrium transport, and the Joule heating is quadratic in the current or voltage). The corresponding balance equation is

$$I_c = A Q_c = \pi_n I - \kappa \frac{A}{L} \Delta T - \frac{I^2 R_n}{2} \qquad \text{W}, \qquad (4.28)$$

where

$$R_n = \rho \frac{L}{A} \qquad (4.29)$$

is the resistance of the n-type leg and $\Delta T = T_H - T_C$. Equation (4.28) shows that Joule heating reduces the magnitude of the heat pumped. To find the maximum heat that can be pumped, we solve $dQ_c/dI = 0$ to find $I_c^{\max}$, which can be inserted in eqn. (4.28) to find the corresponding maximum heat that can be pumped at the cold junction, $Q_c^{\max}$. To determine the maximum temperature difference that can be supported, we set $Q_c^{\max} = 0$ and find

$$\Delta T_{\max} = T_H - T_C = \frac{1}{2} Z T_C^2, \qquad (4.30)$$

where

$$Z = \frac{S_n^2 \sigma_n}{\kappa} \qquad (4.31)$$

is the so-called *thermoelectric figure of merit*, an important quantity. The maximum temperature difference occurs when the current is set to I_c^{max} and $Q_c^{\text{max}} = 0$. Under these conditions, the back flow of heat and Joule heating exactly cancel the Peltier heat pumped at the cold junction.

The next question is what determines the cooling efficiency or coefficient of performance (COP)? The efficiency of the Peltier heat pump is the ratio of the heat pumped to the input electrical power,

$$\eta = \frac{Q_c}{P_{\text{in}}} \, . \tag{4.32}$$

We can calculate the efficiency in two different ways. We could determine the maximum efficiency by evaluating $d\eta/dI = 0$ for the current at the maximum efficiency. Inserting that current in eqn. (4.32) would give the maximum efficiency. Alternatively, we could use the current, I_c^{max}, that gives the maximum heat pumped, insert it in eqn. (4.32) to find the COP at under maximum cooling power conditions. In either case, the answer can be written as [7]

$$\eta = COP = \frac{Q_c}{P_{\text{in}}} = f_P \left(T_H, T_C, Z \right) \, . \tag{4.33}$$

The important point is that the efficiency of the Peltier cooler is given by a function, f_P, of the hot side temperature, cold side temperature, and the thermoelectric figure of merit. Both the maximum temperature difference possible and the COP of the Peltier cooler depend on the thermoelectric parameters in the combination as given by eqn. (4.31). Qualitatively, it is easy to see why. Higher conductivities lower the Joule heating losses, higher Seebeck coefficients increase the Peltier heat pumped, and lower thermal conductivities reduce the back flow of heat from the hot side to the cold side.

A similar calculation can be done for the thermoelectric power generator sketched in Fig. 4.9. Again, we would set up a heat flow balance equation but this time at the hot side junction. The heat flux input (which we are trying to convert into electricity) is equal to the Peltier heat pumped at the hot side, plus the heat that diffuses from the hot to cold junction, minus one-half of the Joule heating. The conversion efficiency is simply the ratio of the output electrical power to the input heat current,

$$\eta = \frac{P_{\text{out}}}{AQ_{\text{in}}} = \frac{I^2 R_L}{AQ_{\text{in}}} \, , \tag{4.34}$$

where R_L is the resistance of the load. The current can be related to the temperature drop from the hot to cold sides. To find the maximum efficiency, we solve $d\eta/dR_L = 0$ to find the optimum load resistance, insert it in eqn. (4.34). As for the Peltier cooler, we find that the power conversion efficiency is determined by the thermoelectric figure of merit, Z.

The most important point of this short discussion is the fact that the efficiency of a thermoelectric device, whether operated as a refrigerator or as a power generator, is determined by the thermoelectric figure of merit, Z, which depends on the properties of the thermoelectric material. In practice, contact and interface resistance (both electrical and thermal) reduce the performance of actual devices, but the material parameter, Z, plays a central role in thermoelectric technology.

4.6 Discussion

Before we conclude this lecture, there are three more things to discuss. The first is the thermoelectric figure of merit (FOM), given its importance in determining device performance. The second is how to think about TE devices in terms of electron flow alone, not electrons and holes as sketched in Figs. 4.8 and 4.9. This is important because our general model for transport developed in Lecture 2, the conductivity derived in Lecture 3, and the thermoelectric parameters developed in this lecture and in the next one are all expressed in terms of electrons. The expressions apply to both n-type and p-type materials, but they refer to *electrons* flowing in the conduction or valence bands respectively. Finally, we briefly discuss the measurement of Seebeck coefficients.

The figure of merit is commonly written as ZT, where

$$ZT = \frac{S_n^2 \sigma_n T_L}{\kappa_n + \kappa_L} \qquad (4.35)$$

is a dimensionless number. A good thermoelectic has a FOM of about one. For widespread applications in electronic cooling or power generation from waste heat, a FOM of ≈ 3 is desired. This is the grand challenge of thermoelectric research. Let's consider two questions: 1) What material properties determine ZT? and 2) Given a material, how do we optimize ZT?

According to eqn. (4.23), the Seebeck coefficient mainly depends on the difference between the band edge and the Fermi level. Details of band structure and scattering affect the parameter, δ_n, but the differences are rather small. According to eqn. (4.22), the conductivity depends on the effective number of channels for conduction, $\langle M \rangle$, and the average mean

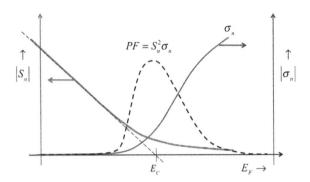

Fig. 4.11.   Sketch of the Seebeck coefficient vs. Fermi level (left axis), conductivity vs. Fermi level (right axis), and power factor, PF (dashed line).

free path for backscattering, $\langle\langle\lambda\rangle\rangle$. To get a large $\langle M\rangle$, we need the Fermi level to be high, preferably in the band, and a large $M(E)$. For a long mean-free-path, scattering should be weak (high mobility). The denominator of the FOM is the thermal conductivity. Typically, $\kappa_L \gg \kappa_n$. In Lecture 9 we will discuss the lattice thermal conductivity.

Figure 4.11 is a sketch of S_n and σ_n vs. the position of the Fermi level. As the Fermi level moves towards and then into the conduction band, $|S_n|$ decreases, but at the same time, σ_n increases because there are more and more channels to conduct. The product of the two is the power factor, $PF = S_n^2 \sigma_n$, which is maximized when E_F is near the band edge. The precise location depends on details of band structure and scattering, but in practice, TE device designers seek to maximize performance by doping the material to place E_F near the band edge.

Figure 4.12 shows the thermoelectric cooler in terms of electron flow. Electrons flow from the top metal, into the n-type leg, and down to the bottom left contact. Electrons flow from the bottom right contact into the p-type leg and up to the top metal contact. The energy band diagrams show how to think about Peltier cooling in terms of electrons only. For example, at the upper left, we see electrons flowing from the metal, and absorbing heat in order to move into the conduction band of the n-type semiconductor. On the upper right, we see electrons moving up in the valence band of the p-type leg absorbing energy to occupy an empty state in the top metal. At the bottom left, we see that heat is emitted when electrons move from the conduction band of the n-type leg to the metal. As shown on the lower right, electrons that drop down in energy from the

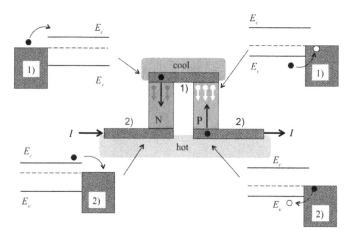

Fig. 4.12.   Illustration of Peltier cooling operation in terms of electron flow alone, rather than electrons and holes.

metal to fill an empty state in the valence band of the p-type semiconductor, also emit heat. For either an n-type or p-type semiconductor, we can view conduction in terms of electrons or holes, which ever is more convenient.

Finally, let's discuss how one would measure the Seebeck coefficient. Figure 4.13 is a simple illustration. Two contacts are made to the sample, and a heater on one side introduces a temperature gradient. To determine the temperature difference across the sample, we need to measure the temperature at two contacts. In the illustration, this is done with two thermocouples (which also operate on the Seebeck effect). (In practice, some care is needed to accurately measure the temperature gradient.) A high impedance voltmeter is attached to measure the open-circuit voltage.

The Seebeck coefficient of the sample is

$$S_s = \frac{-\Delta V_s}{\Delta T},\tag{4.36}$$

but the voltmeter does not measure only the voltage drop across the sample. The leads of the voltmeter are attached to the contacts. For the lead at the right, one end is at contact 2 and, therefore, at temperature, T_{L2} while the other end is at the voltmeter. If the temperature of the voltmeter is T_{L1}, then there is a temperature difference and therefore a Seebeck voltage across this lead with a value of

$$S_l = \frac{-\Delta V_l}{\Delta T}.\tag{4.37}$$

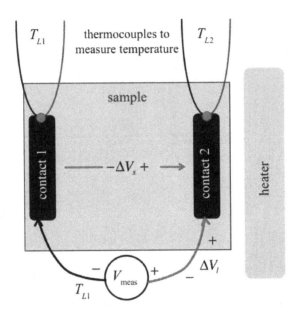

Fig. 4.13.   Illustration of how the Seebeck coefficient is measured.

Note that this voltage opposes the sample Seebeck voltage. Kirchoff's Voltage Law tells us that

$$V_{\mathrm{meas}} = \Delta V_s - \Delta V_l = - (S_s - S_l) \Delta T, \qquad (4.38)$$

so the measured voltage gives the difference between the Seebeck coefficient of the sample and that of the lead. The leads are typically gold or copper and have a low Seebeck coefficient, but when measuring small Seebeck coefficients, the lead correction can be important. It is also interesting to think about how eqn. (4.15) would change if we take the Peltier coefficient of the metal contacts into account.

4.7   Summary

Our goal in this lecture has been to understand how temperature gradients affect electrical currents and how electrical currents affect the flow of heat. In Lecture 3, we focused on eqn. (3.2), which applies quite generally when the temperature is uniform. In this lecture, we used heuristic arguments to develop the basic general equations of thermoelectricity, eqns. (4.21). We have assumed a 3D conductor and diffusive transport, but in Lecture 5 we will see how to formally derive similar equations in 1D, 2D, or 3D and from ballistic to diffusive conditions.

In summary, in this lecture, we have:

(1) Discussed the physics of thermoelectricity — the Seebeck and Peltier effects.
(2) Developed the basic equations that describe thermoelectricity.
(3) Discussed how the four thermoelectric parameters depend on the properties of the material.
(4) Introduced thermoelectric devices and the thermoelectric figure of merit, ZT, which controls their performance.

Having discussed the physics, we turn in the next lecture to a formal, mathematical treatment of thermoelectricity.

4.8 References

The approach used in this lecture is similar to that presented by Datta:

[1] Supriyo Datta, *Lessons from Nanoelectronics: A new approach to transport theory*, World Scientific Publishing Company, Singapore, 2011.

For the conventional approach to thermoelectric transport, see:

[2] N.W. Ashcroft and N.D. Mermin, *Solid–State Physics*, Saunders College, Philadelphia, PA, 1976.

Smith, Janek, and Adler provide a good discussion of the Onsaeger relations for coupled flows.

[3] A.C. Smith, J. Janak, and R. Adler, *Electronic Conduction in Solids*, McGraw-Hill, New York, NY 1965.

For a brief treatment of semiconductor fundamentals, see:

[4] Robert F. Pierret, *Advanced Semiconductor Fundamentals, 2nd Ed.*, Vol. VI, Modular Series on Solid-State Devices, Prentice Hall, Upper Saddle River, N.J., USA, 2003.

For the pioneering book on thermoelectric technology, see

[5] A.F. Ioffe, *Semiconductor Thermoelements and Thermoelectric Cooling*, Infosearch, London, 1957.

In recent years, many ideas to make use of nanotechnology to enhance thermoelectric device efficiency have been proposed. For a recent review, consult:

[6] A. Majumdar, "Thermoelectricity in semiconductor nanostructures", *Science*, **303**, 778-779, 2004.

[7] M. Dresselhaus, G. Chen, M. Tang, R. Yang, H. Lee, D. Wang, Z. Ren, J.-P. Fleurial, and P. Gogna, "New Directions for low dimensional thermoelectric materials", *Advanced Materials*, **19**, 1043-1053, 2007.

[8] A.J. Minnich, M.S. Dresselhaus, Z.F. Ren, and G. Chen, "Bulk nanostructured thermoelectric materials: current research and future prospects", *Energy and Environmental Science*, **2**, 466-479, 2009.

The design of thermoelectric devices is discussed by Hode:

[9] Marc Hode, "On One-Dimensional Analysis of Thermoelectric Modules (TEMs)", *IEEE Trans. on Components and Packaging Technologies*, **28**, 218-229, 2005.

Marc Hode, "Optimal Pellet Geometries for Thermoelectric Refrigeration", *IEEE Trans. on Components and Packaging Technologies*, **30**, 50-58, 2007.

Marc Hode, "Optimal Pellet Geometries for Thermoelectric Power Generation", *IEEE Trans. on Components and Packaging Technologies*, **33**, 307-318, 2010.

Lecture 5

Thermoelectric Effects: Mathematics

Contents

5.1   Introduction

In Lecture 4, we discussed the physics of thermoelectricity using heuristic mathematical arguments to develop the basic equations. Our discussion assumed a 3D conductor and diffusive transport. You might wonder how to describe thermoelectricity in 1D or 2D and under ballistic or quasi-ballistic conditions. To answer these questions, we need a formal, mathematical description of thermoelectricity. We present a derivation of the basic equations of thermoelectricity in this lecture.

In Lecture 2 we saw that the electrical current for our generic device is

$$I = -I_x = \frac{2q}{h} \int T(E)M(E)\,(f_1 - f_2)\,dE\,. \tag{5.1}$$

Anything that causes a difference between f_1 and f_2 causes current to flow. (The minus sign in this expression reminds us that the current was defined to be positive when it flows *into* contact 2, so positive current flows in the $-x$ direction.) Differences in the Fermi levels of the two contacts (caused by an applied voltage across the device) and differences in the temperatures of the two contacts can both cause f_1 to be different from f_2. The two driving forces for current flow are differences in the Fermi levels and temperatures of the two contacts.

Electrons are particles that carry both charge and heat. The charge current is given by eqn. (5.1). To get the heat current, we note that electrons in the contact flow at an energy, $E \approx E_F$. To enter an energy channel, E, in the device, electrons must absorb (if $E > E_F$) or emit (if $E < E_F$) a thermal energy of $E - E_F$. We conclude that for the heat current term, we should replace the q in eqn. (5.1) with $(E - E_F)$ and move it inside the integral. The resulting heat current is

$$I_Q = \frac{2}{h} \int (E - E_F)\, T(E) M(E)\, (f_1 - f_2)\, dE. \tag{5.2}$$

Thermoelectricity involves near-equilibrium transport where $f_1 \approx f_2 \approx f_0$. In the next section, we develop a Taylor series expansion of $(f_1 - f_2)$ for small differences in voltage and temperature.

5.2 Driving forces for current flow

Figure 5.1(a) is a sketch of the Fermi function for two different locations of the Fermi level. When $\Delta V = V_2 - V_1 > 0$, we see that $f_1 > f_2$ for all energies, so the total current is positive. The sign of the current does not depend on whether the semiconductor is n-type or p-type. We saw in Lecture 2 that for small differences in the two Fermi levels, we can expand $(f_1 - f_2)$ in a Taylor series. Keeping only the first term, we found

$$(f_1 - f_2) \approx \left(-\frac{\partial f_0}{\partial E} \right) q\Delta V. \tag{5.3}$$

The term, $(-\partial f_0/\partial E)$ is called the *window function*; it gives the range of energies that contribute to current flow.

Figure 5.1(b) is a sketch of the Fermi function for the case in which the two Fermi levels are identical, but the two temperatures are different, $\Delta T_L = T_{L2} - T_{L1} > 0$. We see that $f_1 > f_2$ for energies below E_F, and $f_1 < f_2$ for energies above E_F. Current flows, but the sign of the current depends on whether the channels are located above E_F (n-type) or below (p-type). For near-equilibrium transport, we can expand $(f_1 - f_2)$ as

$$(f_1 - f_2) \approx f_1 - \left(f_1 + \frac{\partial f_1}{\partial T_L} \Delta T \right) = -\frac{\partial f_0}{\partial T_L} \Delta T. \tag{5.4}$$

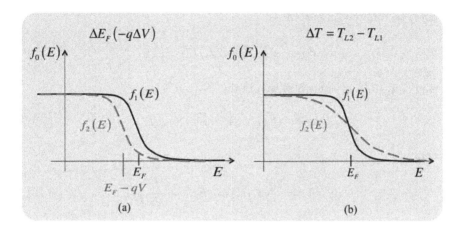

Fig. 5.1. Sketch of the Fermi functions of the two contacts when: (a) the two voltages are different but the temperatures are the same and (b) the two voltages are identical but the temperatures are different.

Differentiating the Fermi function, we find

$$(f_1 - f_2) \approx - \left(-\frac{\partial f_0}{\partial E} \right) \frac{(E - E_F)}{T_L} \Delta T \,. \tag{5.5}$$

In general, there can be differences in both the Fermi levels and the temperatures of the two contacts. The total difference in f_1 and f_2 is the sum of eqns. (5.3) and (5.5):

$$\boxed{(f_1 - f_2) \approx \left(-\frac{\partial f_0}{\partial E} \right) q\Delta V - \left(-\frac{\partial f_0}{\partial E} \right) \frac{(E - E_F)}{T_L} \Delta T \,.} \tag{5.6}$$

The two driving forces for current flow are related to differences in Fermi level (or voltage) and differences in temperature, and for small differences, they simply add.

5.3 Charge current

Deriving a general, near-equilibrium current equation is now straightforward. The total current is the sum of the contributions from each energy channel,

$$I = \int I'(E) \, dE \,, \tag{5.7}$$

where the differential current is

$$I'(E) = \frac{2q}{h}T(E)M(E)(f_1 - f_2).$$ (5.8)

Using eqn. (5.6) in eqn. (5.8) we obtain

$$I'(E) = G'(E)\Delta V + S'_T(E)\Delta T,$$ (5.9)

where

$$G'(E) = \frac{2q^2}{h}T(E)M(E)\left(-\frac{\partial f_0}{\partial E}\right)$$ (5.10)

is the differential conductance and

$$S'_T(E) = -\frac{2q^2}{h}T(E)M(E)\left(\frac{E - E_F}{qT_L}\right)\left(-\frac{\partial f_0}{\partial E}\right)$$

$$= -\left(\frac{k_B}{q}\right)\left(\frac{E - E_F}{k_B T_L}\right)G'(E),$$ (5.11)

is related to the Soret coefficient for electro-thermal diffusion. Note that $S'_T(E)$ is negative for channels above E_F and positive for channels below E_F.

To find the total charge current, we integrate eqn. (5.9) over the energy channels to find

$$I = G\Delta V + S_T\Delta T,$$ (5.12)

where

$$G = \int G'(E)dE,$$ (5.13)

and

$$S_T = \int S'_T(E)dE$$ (5.14)

with the differential conductance being given by eqn. (5.10) and the differential Soret coefficient by eqn. (5.11). These equations are valid in 1D, 2D, or 3D and from the ballistic to diffusive limits.

Exercise: Current equation for bulk transport

In Lecture 2, we saw that the current equation that describes near-equilibrium transport in a bulk material with a uniform temperature is

$$J_{nx} = \sigma_n \frac{d\,(F_n/q)}{dx}$$

$$J_{nx} = \sigma_n \mathcal{E}_x\,. \tag{5.15}$$

The second form of this equation applies when the carrier density is uniform, and there are no diffusion coefficients. How does this equation change in the presence of a temperature gradient?

We begin with eqn. (5.12) and recall that in eqn. (5.12) a positive current flows in the $-x$ direction. Dividing by $-A$ to find the current density in the $+x$ direction, we find

$$J_{nx} = -\frac{G}{A}\Delta V - \frac{S_T}{A}\Delta T\,. \tag{5.16}$$

If we multiply and divide by the length of the resistor, this becomes

$$J_{nx} = -G\frac{L}{A}\frac{\Delta V}{L} - S_T\frac{L}{A}\frac{\Delta T}{L}\,. \tag{5.17}$$

In the diffusive limit, $G = \sigma_n A/L$ and $S_T = s_T A/L$. We also have $dF_n/dx \approx -q\Delta V/L$ and $dT_L/dx \approx \Delta T/L$, so eqn. (5.17) becomes

$$\boxed{J_{nx} = \sigma_n \frac{d(F_n/q)}{dx} - s_T \frac{dT_L}{dx}\,,} \tag{5.18}$$

which is the proper current equation in the bulk when there are gradients in both the electrochemical potential and the temperature.

Please note the notation in the above equations. The subscript "n" suggests that we are describing an n-type conductor. In that case, σ_n is positive and s_T is negative. But our starting point, eqn. (5.1), describes electron flow in *either* the conduction or valence bands. For a p-type semiconductor, the current flow is by electrons in the valence band. The subscripts would change to "p"; σ_p would be positive, but in this case, s_T would also be positive. We'll discuss more about n- and p-type conduction in Sec. 5.5.

5.4   Heat current

Figure 5.2 is an illustration of heat flow in our generic device. Heat is absorbed at contact 1 and emitted at contact 2. Following eqn. (5.2), we

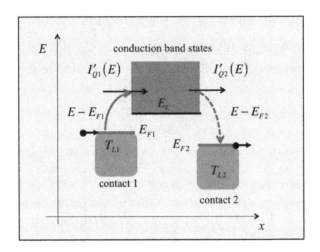

Fig. 5.2.   Schematic illustration of heat absorption and emission in the generic device.

write the heat fluxes at contacts 1 and 2 according to

$$I_{Q1} = \frac{2}{h} \int (E - E_{F1}) \, T(E) M(E) \, (f_1 - f_2) \, dE$$

$$I_{Q2} = \frac{2}{h} \int (E - E_{F2}) \, T(E) M(E) \, (f_1 - f_2) \, dE \,.$$

(5.19)

Assuming near-equilibrium conditions and using eqn. (5.6), we find

$$I'_Q(E) = -T_L S'_T(E) \Delta V - K'_0(E) \Delta T \,,$$

(5.20)

where we have suppressed the label for the contact numbers and

$$K'_0(E) = \frac{(E - E_{F1})^2}{q^2 T_L} G'(E) \,.$$

(5.21)

To find the total heat current, we integrate over all of the energy channels and find

$$I_Q = -T_L S_T \Delta V - K_0 \Delta T \,,$$

(5.22)

where

$$K_0 = T_L \left(\frac{k_B}{q}\right)^2 \int \left(\frac{E - E_{F1}}{k_B T_L}\right)^2 G'(E) dE$$

(5.23)

is the electronic heat conductance under short circuit ($\Delta V = 0$) conditions.

5.5   Discussion

Beginning with the general equations, we have derived the charge and heat current equations for near-equilibrium transport as

$$I = G\Delta V + S_T \Delta T$$
$$I_Q = -T_L S_T \Delta V - K_0 \Delta T,$$

(5.24)

which shows that differences in voltage and temperature cause both charge and heat currents to flow. The units of I are Amperes and I_Q Watts. The conductance, G, is in Siemans (1/Ohms), S_T in Amperes/Kelvin, and K_0 is Watts/Kelvin. The general expressions for the three transport parameters are

$$G'(E) = \frac{2q^2}{h} T(E) M(E) \left(-\frac{\partial f_0}{\partial E} \right)$$

$$G = \int G'(E) dE$$

$$S_T = -\left(\frac{k_B}{q} \right) \int \left(\frac{E - E_F}{k_B T_L} \right) G'(E) dE$$

$$K_0 = T_L \left(\frac{k_B}{q} \right)^2 \int \left(\frac{E - E_F}{k_B T_L} \right)^2 G'(E) dE.$$

(5.25)

(Recall, again, that K_0 describes only the part of heat conduction due to electrons. In semiconductors, the larger contribution comes from the lattice.) Equations (5.25) are valid in 1D, 2D, or 3D and for ballistic to diffusive transport. For 3D, diffusive transport, the transport equations become

$$\boxed{\begin{aligned} J_{nx} &= \sigma_n \frac{d(F_n/q)}{dx} - s_T \frac{dT_L}{dx} \\ J_{Qx} &= T_L s_T \frac{d(F_n/q)}{dx} - \kappa_0 \frac{dT_L}{dx}. \end{aligned}}$$

(5.26)

These equations have the same form in 1D and 2D, but the units of the various terms differ. The equations are often written with $d(F_n/q)/dx$ replaced by $\mathcal{E}_x$, which is the electric field. This replacement is valid when the carrier density is uniform, and diffusion currents are absent. In eqns. (5.26), J_{nx} is in A/m^2 and J_{Qx} is in (W/m^2). The conductivity, σ_n, has units of 1/Ω-m, s_T has units of A/m-K, and κ_0 has units of W/m-K. The thermoelectric transport parameters for 3D, diffusive transport are

$$
\sigma'_n (E) = \frac{2q^2}{h} M_{3D}(E) \lambda(E) \left(-\frac{\partial f_0}{\partial E} \right)
$$

$$
\sigma_n = \int \sigma'_n(E)dE
$$

$$
s_T = - \left(\frac{k_B}{q} \right) \int \left(\frac{E - E_F}{k_B T_L} \right) \sigma'_n(E)dE
$$

$$
\kappa_0 = T_L \left(\frac{k_B}{q} \right)^2 \int \left(\frac{E - E_F}{k_B T_L} \right)^2 \sigma'_n(E)dE \,.
$$

(5.27)

Recall that $M_{3D}(E) = M(E)/A$.

Inverted form of the transport equations

Equations (5.24) and (5.26) are in the form that results naturally from our generic model for current flow, eqns. (5.1) and (5.2). They correspond experimentally to a situation in which voltage and temperature differences are applied and the charge and heat currents that flow are measured. (Voltage and temperature are the independent quantities and the currents are the dependent quantities.) In this form of the equations, the contributions from each energy channel add. For experiments, however, it is often convenient to re-write these equations so that the charge current and temperature differences are the independent quantities. Accordingly, eqns. (5.24) become

$$
\Delta V = RI - S\Delta T
$$

$$
I_Q = -\Pi I - K_n \Delta T \,,
$$

(5.28)

where

$$
S = \frac{s_T}{G}
$$

$$
\Pi = T_L S
$$

(5.29)

$$
K_n = K_0 - \Pi S G \,.
$$

Note that in this form of the equations, the contributions from each energy channel do not add (i.e. $R \neq \int R(E)dE$).

For 3D, diffusive transport, the transport equations in the inverted form become

$$\frac{d\left(F_n/q\right)}{dx} = \rho_n J_{nx} + S_n \frac{dT_L}{dx}$$

$$J_{Qx} = T_L S_n J_{nx} - \kappa_n \frac{dT_L}{dx} ,$$

(5.30)

which should be compared to eqns. (5.26). The transport parameters in eqns. (5.30) are

$$\rho_n = 1/\sigma_n$$
$$S_n = s_T/\sigma_n$$
$$\kappa_n = \kappa_0 - S_n^2 \sigma_n T_L ,$$

(5.31)

which correspond to those in eqns. (5.27) for the transport eqns. (5.26). Again, one frequently sees eqn. (5.30) written with $d(F_n/q)/dx$ replaced by $\mathcal{E}_x$.

Exercise 5.1: Transport parameters in 1D

As an example of how to apply the general relations developed in this lecture, let's work out the transport coefficients for a specific case. Consider a 1D conductor in the ballistic limit with one subband occupied, and let's evaluate the transport parameters in eqns. (5.24) or (5.28). In this case, $T(E) = 1$ and $M(E) = g_v$, where g_v is the valley degeneracy. We assume that only one subband is occupied, but depending on the band structure, there may be more than one degenerate valley. For example, in a carbon nanotube, $g_v = 2$ [1]. Note that in 1D, the valley degeneracy is the only information about band structure that we need.

To evaluate the conductance (or resistance), we begin with the differential conductance, eqn. (5.10), and find

$$G'\left(E\right) = \frac{2q^2}{h} g_v \left(-\frac{\partial f_0}{\partial E}\right) ,$$

(5.32)

so the conductance becomes

$$G = \frac{1}{R} = \frac{2q^2}{h} g_v \int_{E_c}^{\infty} \left(-\frac{\partial f_0}{\partial E}\right) dE = \frac{2q^2}{h} \langle M \rangle .$$

(5.33)

The integral is the number of channels for conduction, which for fully degenerate conditions ($T_L = 0$ K) is $\langle M \rangle = g_v$. In general, however, only a fraction of the channel are occupied. The integral is

$$\langle M \rangle = g_v \int_{E_c}^{\infty} \left(-\frac{\partial f_0}{\partial E} \right) dE = g_v \frac{\partial}{\partial E_F} \int_{E_c}^{\infty} f_0 \, dE \,, \qquad (5.34)$$

where we have used $(-\partial f_0 / \partial E) = (+\partial f_0 / \partial E_F)$, which comes from the form of the Fermi function.

To find $\langle M \rangle$, we must evaluate

$$\langle M \rangle = g_v \frac{\partial}{\partial E_F} \int_{E_c}^{\infty} \frac{dE}{1 + e^{(E - E_F)/k_B T_L}} \,, \qquad (5.35)$$

which can be done by defining the variables, $\eta = (E - E_c)/k_B T_L$ and $\eta_F = (E_F - E_c)/k_B T_L$ so that eqn. (5.35) becomes

$$\langle M \rangle = g_v \frac{\partial}{\partial \eta_F} \int_0^{\infty} \frac{d\eta}{1 + e^{\eta - \eta_F}} \,. \qquad (5.36)$$

The integral can be recognized as the Fermi-Dirac integral of order 0. Using the rule for differentiating Fermi-Dirc integrals, eqn. (3.24), we finally obtain

$$\begin{aligned} \langle M \rangle &= g_v \mathcal{F}_{-1} (\eta_F) \\ G &= \frac{2q^2}{h} \langle M \rangle = \frac{2q^2}{h} g_v \mathcal{F}_{-1} (\eta_F) \,. \end{aligned} \qquad (5.37)$$

Before we compute the other parameters, let's examine the result. For a non-degenerate material, $\eta_F \ll 0$ and $\mathcal{F}_{-1}(\eta_F) \to e^{\eta_F}$. For a strongly degenerate material, $\eta_F \gg 0$. The Fermi-Dirac integral, $\mathcal{F}_{-1}(\eta_F)$ is an analytical function [2],

$$\mathcal{F}_{-1}(\eta_F) = \frac{\partial \mathcal{F}_0}{\partial \eta_F} = \frac{\partial \left[\ln \left(1 + e^{\eta_F} \right) \right]}{\partial \eta_F} = \frac{e^{\eta_F}}{1 + e^{\eta_F}} \,. \qquad (5.38)$$

For $\eta_F \gg 0$, we see that $\mathcal{F}_{-1}(\eta_F) \to 1$ and $G = (2q^2/h) \, g_v$, as expected.

Having computed G and $R = 1/G$ for the 1D ballistic conductor, we now turn to the Soret and Seebeck coefficients. From eqn. (5.25) we find

$$S_T = - \left(\frac{k_B}{q} \right) \int_{E_c}^{\infty} \left(\frac{E - E_F}{k_B T_L} \right) \left[\frac{2q^2}{h} g_v \left(-\frac{\partial f_0}{\partial E} \right) \right] dE \,. \qquad (5.39)$$

Let's multiply and divide this by G as given by eqn. (5.33). Doing so, we find

$$S_T = - \left(\frac{k_B}{q} \right) \frac{\int_{E_c}^{\infty} \left(\frac{E - E_F}{k_B T_L} \right) \left(-\frac{\partial f_0}{\partial E} \right) dE}{\int_{E_c}^{\infty} \left(-\frac{\partial f_0}{\partial E} \right) dE} G \,, \qquad (5.40)$$

so S_T is proportional to G, and we only need to evaluate the two integrals. We have seen already that the denominator is $\mathcal{F}_{-1}(\eta_F)$, so we only need to evaluate the numerator.

Working on the numerator of eqn. (5.40), we find

$$
\begin{aligned}
\text{num} &= \int_{E_c}^{\infty} \left(\frac{E - E_F}{k_B T_L} \right) \left(-\frac{\partial f_0}{\partial E} \right) dE \\
&= \int_{E_c}^{\infty} \left(\frac{E - E_c + E_c - E_F}{k_B T_L} \right) \left(-\frac{\partial f_0}{\partial E} \right) dE \\
&= \int_{E_c}^{\infty} \left(\frac{E - E_c}{k_B T_L} \right) \left(+\frac{\partial f_0}{\partial E_F} \right) dE - \eta_F \int_{E_c}^{\infty} \left(+\frac{\partial f_0}{\partial E_F} \right) dE .
\end{aligned}
\tag{5.41}
$$

Now proceeding as before, we move $\partial/\partial E_F$ outside of the integrals and change variables to $\eta = (E - E_c)/k_B T_L$ and $\eta_F = (E_F - E_c)/k_B T_L$ and find

$$
\text{num} = \frac{\partial}{\partial \eta_F} \int_0^{\infty} \frac{\eta \, d\eta}{1 + e^{\eta - \eta_F}} - \eta_F \frac{\partial}{\partial \eta_F} \int_0^{\infty} \frac{d\eta}{1 + e^{\eta - \eta_F}} .
\tag{5.42}
$$

We recognize the first integral as $\mathcal{F}_1(\eta_F)$, and the differentiation gives us $\mathcal{F}_0(\eta_F)$. We recognize the second integral as $\mathcal{F}_0(\eta_F)$ and the differentiation gives us $\mathcal{F}_{-1}(\eta_F)$, so the numerator of eqn. (5.40) becomes

$$
\text{num} = - \left(\frac{k_B}{q} \right) [\mathcal{F}_0(\eta_F) - \eta_F \mathcal{F}_{-1}(\eta_F)] ,
\tag{5.43}
$$

and S_T becomes

$$
S_T = - \left(\frac{k_B}{q} \right) \left(-\eta_F + \frac{\mathcal{F}_0(\eta_F)}{\mathcal{F}_{-1}(\eta_F)} \right) G .
\tag{5.44}
$$

The Seebeck coefficient, $S = S_T/G$, is

$$
S = - \left(\frac{k_B}{q} \right) \left(-\eta_F + \frac{\mathcal{F}_0(\eta_F)}{\mathcal{F}_{-1}(\eta_F)} \right) ,
\tag{5.45}
$$

which, recalling the definition of $\eta_F = (E_F - E_c)/k_B T_L$, can be written as

$$
\begin{aligned}
S_n &= - \left(\frac{k_B}{q} \right) \left(\frac{E_c - E_F}{k_B T_L} + \delta_n \right) \\
\delta_n &= \frac{\mathcal{F}_0(\eta_F)}{\mathcal{F}_{-1}(\eta_F)} .
\end{aligned}
\tag{5.46}
$$

Before proceeding to the evaluation of electronic heat conductance, let's examine the Seebeck coefficient. First, we see that from eqn. (5.46) that S_n

has the expected form of eqn. (4.23). Second, in the non-degenerate limit, the Fermi-Dirac integrals become exponentials, and $\delta_n \to 1$, or Δ_n becomes $k_B T_L$. For a non-degenerate, 1D ballistic conductor, the current flows on average, at an energy, $k_B T_L$, above the bottom of the band. Finally, for strongly degenerate conductors, $\eta_F \gg 0$ and from the expression for $\mathcal{F}_0$ and $\mathcal{F}_{-1}$ (see eqn. (5.38)), we see that $\mathcal{F}_0 / \mathcal{F}_{-1} \to \eta_F$. For strongly degenerate conditions, $\delta_n \to \eta_F = (E_F - E_c)/k_B T_L$, and eqn. (5.46) shows that $S_n \to 0$. For strong carrier degeneracy, $\Delta_n \to (E_F - E_c)$, which is much greater than the $k_B T_L$ value under non-degenerate conditions.

We have computed G and $R = 1/G$ as well as S_T, S_n, and $\Pi = T_L S_n$ for the 1D ballistic conductor. We now turn to the electronic heat conductances, K_0 and K_n. From eqn. (5.25), after dividing and multiplying by the conductance, we find

$$K_0 = T_L \left(\frac{k_B}{q} \right)^2 \left[\frac{\int_{E_c}^{\infty} \left(\frac{E - E_F}{k_B T_L} \right)^2 \left(-\frac{\partial f_0}{\partial E} \right) dE}{\int_{E_c}^{\infty} \left(-\frac{\partial f_0}{\partial E} \right) dE} \right] G . \qquad (5.47)$$

We see that the electronic thermal conductance is proportional to the electrical conductance, which is just a statement of the Wiedemann-Franz Law, which we will discuss next. Again, we recognize the denominator as $\mathcal{F}_{-1}(\eta_F)$, so we only need to evaluate the numerator. Working on the numerator of eqn. (5.47), we find

$$\text{num} = \int_{E_c}^{\infty} \left(\frac{E - E_F}{k_B T_L} \right)^2 \left(-\frac{\partial f_0}{\partial E} \right) dE$$

$$= \int_{E_c}^{\infty} \left[\frac{(E - E_c) + (E_c - E_F)}{k_B T_L} \right]^2 \left(-\frac{\partial f_0}{\partial E} \right) dE$$

$$= \int_{E_c}^{\infty} \left(\frac{E - E_c}{k_B T_L} \right)^2 \left(+\frac{\partial f_0}{\partial E_F} \right) dE - 2\eta_F \int_{E_c}^{\infty} \left(\frac{E - E_c}{k_B T_L} \right) \left(+\frac{\partial f_0}{\partial E_F} \right) dE$$

$$+ \eta_F^2 \int_{E_c}^{\infty} \left(+\frac{\partial f_0}{\partial E_F} \right) dE . \qquad (5.48)$$

The last term in eqn. (5.48) can be recognized as $\eta_F^2 \mathcal{F}_{-1}$, and the second term is $-2\eta_F \mathcal{F}_0$. Finally, the first term is $2\mathcal{F}_1$. Putting this all together, we find

$$K_0 = T_L \left(\frac{k_B}{q} \right)^2 \left[2 \frac{\mathcal{F}_1(\eta_F)}{\mathcal{F}_{-1}(\eta_F)} - 2\eta_F \frac{\mathcal{F}_0(\eta_F)}{\mathcal{F}_{-1}(\eta_F)} + \eta_F^2 \right] G . \qquad (5.49)$$

Equation (5.49) gives the electronic heat conductance under short circuit conditions. To find the open circuit electronic heat conductance, K_n, we use the definition, eqn. (5.29) and the result for S_n, eqn. (5.45), to find

$$K_n = T_L \left(\frac{k_B}{q}\right)^2 \left[2\frac{\mathcal{F}_1(\eta_F)}{\mathcal{F}_{-1}(\eta_F)} - \left(\frac{\mathcal{F}_0(\eta_F)}{\mathcal{F}_{-1}(\eta_F)}\right)^2\right] G. \qquad (5.50)$$

The point of this exercise has been to illustrate how the general expressions we have developed are evaluated in specific cases. The procedure is straight-forward, but it can get tedious. Depending on the application, it may be preferable to just numerically integrate the expressions.

For additional practice, consider diffusive transport in this 1D conductor $(T(E) = \lambda(E)/L)$ with power law scattering $(\lambda(E) = \lambda_0[(E-E_c)/k_B T_L]^r)$. Your final expressions will contain the characteristic exponent, r, which, as we will discuss in Lecture 6, is typically between 0 and 2. In particular, derive an expression for the parameter, δ_n in terms of r and explain physically why δ_n increases as r increases.

Wiedemann-Franz Law and Lorenz Number

In Lecture 4, we asserted that there should be a relation between the electrical and thermal conductivities because electrons carry both the electrical and thermal currents. We can now make this relation explicit. First, let's define the conductivity-weighted average of a quantity as

$$\overline{(\bullet)} \equiv \frac{\int (\bullet)\sigma'_n(E)dE}{\int \sigma'_n(E)dE}. \qquad (5.51)$$

Using this definition, we can express the transport coefficients, eqns. (5.27) and (5.31) as

$$s_T = -\left(\frac{k_B}{q}\right)\overline{\left(\frac{E-E_F}{k_B T_L}\right)}\sigma_n$$

$$S_n = s_T/\sigma_n = -\left(\frac{k_B}{q}\right)\overline{\left(\frac{E-E_F}{k_B T_L}\right)} = -\left(\frac{k_B}{q}\right)\frac{E_J - E_F}{k_B T_L}$$

$$\kappa_0 = T_L\left(\frac{k_B}{q}\right)^2\overline{\left(\frac{E-E_F}{k_B T_L}\right)^2}\sigma_n \qquad (5.52)$$

$$\kappa_n = T_L\left(\frac{k_B}{q}\right)^2\left[\overline{\left(\frac{E-E_F}{k_B T_L}\right)^2} - \overline{\left(\frac{E-E_F}{k_B T_L}\right)}^2\right]\sigma_n.$$

Equations (5.52) show that the Seebeck coefficient is related to the average energy of current transport, E_J, and that both thermal conductivities, κ_0 and κ_n are proportional to the electrical conductivity, σ_n. From these expressions, we find

$$\frac{\kappa_n}{T_L \sigma_n} = L = \left(\frac{k_B}{q}\right)^2 \left[\overline{\left(\frac{E - E_F}{k_B T_L}\right)^2} - \overline{\left(\frac{E - E_F}{k_B T_L}\right)}^2\right], \tag{5.53}$$

which is just the Wiedemann-Franz Law with L being the Lorenz number. The factor in brackets depends on the shape of the band, the degree of degeneracy, and the type of scattering. For a parabolic energy band and energy independent scattering, this factor is 2 for non-degenerate conditions and $\pi^2/3$ for degenerate conditions.

P-type semiconductors and bipolar conduction

Consider a 3D semiconductor with parabolic energy bands. For the conduction band,

$$M^c_{3D}(E) = g_v \frac{m^*_n}{2\pi\hbar^2} (E - E_c) \qquad E \geq E_c \tag{5.54}$$

and for the valence band

$$M^v_{3D}(E) = g_v \frac{m^*_p}{2\pi\hbar^2} (E_v - E) \qquad E \leq E_v. \tag{5.55}$$

Figure 5.3 is a sketch of $M(E)$ for this simple bandstructure.

The conductivity consists of two parts. For the conduction band,

$$\sigma_n = \frac{2q^2}{h} \int_{E_c}^{\infty} M^c_{3D}(E) \lambda_n(E) \left(-\frac{\partial f_0}{\partial E}\right) dE, \tag{5.56}$$

and for the valence band

$$\sigma_p = \frac{2q^2}{h} \int_{-\infty}^{E_v} M^v_{3D}(E) \lambda_p(E) \left(-\frac{\partial f_0}{\partial E}\right) dE. \tag{5.57}$$

We have not worried about integrating to the top of the conduction band or from the bottom of the valence band but have taken the limits as $\pm\infty$ because the Fermi function ensures that the integrand falls exponentially to zero away from the band edge. The important point is that in both cases we integrate the *same* expression (with the appropriate M_{3D} and λ) over the relevant band. Electrons carry the current in either case, so our general expression is the same for the conduction and valence bands. There

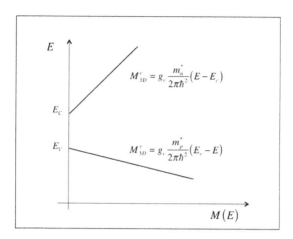

Fig. 5.3. Sketch of $M(E)$ vs. E for a 3D, parabolic band semiconductor.

is no need to change signs for the valence band or to replace $f_0(E)$ with $1 - f_0(E)$.

To compute the Seebeck coefficient for electrons in the conduction band, we have from eqn. (5.27)

$$\sigma'_n(E) = \frac{2q^2}{h} M^c_{3D}(E - E_c)\lambda_n(E)\left(-\frac{\partial f_0}{\partial E}\right)$$

$$\sigma_n = \int_{E_c}^{\infty} \sigma'_n(E)dE$$

$$s_T = -\left(\frac{k_B}{q}\right)\int_{E_c}^{\infty}\left(\frac{E - E_F}{k_B T_L}\right)\sigma'_n(E)dE \qquad (5.58)$$

$$S_n = s_T/\sigma_n = -\left(\frac{k_B}{q}\right)\frac{\int_{E_c}^{\infty}\left(\frac{E-E_F}{k_B T_L}\right)\sigma'_n(E)dE}{\sigma_n}.$$

Similarly, for the Seebeck coefficient for electrons in the valence band is

$$\sigma'_p(E) = \frac{2q^2}{h} M^v_{3D}(E_v - E)\lambda_p(E)\left(-\frac{\partial f_0}{\partial E}\right)$$

$$\sigma_p = \int_{-\infty}^{E_v} \sigma'_p(E)dE$$

$$s_T = -\left(\frac{k_B}{q}\right)\int_{-\infty}^{E_v}\left(\frac{E - E_F}{k_B T_L}\right)\sigma'_p(E)dE \qquad (5.59)$$

$$S_p = s_T/\sigma_p = -\left(\frac{k_B}{q}\right)\frac{\int_{-\infty}^{E_v}\left(\frac{E-E_F}{k_B T_L}\right)\sigma'_p(E)dE}{\sigma_p}.$$

Note that the sign of S_p will be positive.

Finally, we ask the question: What happens if both the conduction and valence bands contribute to conduction? This can occur for a small bandgap or at high temperatures. In this case, we simply integrate over all the channels and find

$$\sigma'(E) = \frac{2q^2}{h} M_{3D}(E) \lambda(E) \left(-\frac{\partial f_0}{\partial E} \right)$$

$$M_{3D}(E) = M_{3D}^v(E) + M_{3D}^c(E) \qquad (5.60)$$

$$\sigma_{\text{tot}} = \int_{-\infty}^{\infty} \sigma'(E) dE = \sigma_n + \sigma_p \,,$$

but what about the Seebeck coefficient? To evaluate S when two bands contribute, remember that in the first form of the transport coefficients, the contributions from each energy channel add in parallel, so the total Soret coefficient is

$$s_T = -\left(\frac{k_B}{q} \right) \int_{-\infty}^{\infty} \left(\frac{E - E_F}{k_B T_L} \right) \sigma'(E) dE = S_n \sigma_n + S_p \sigma_p \,. \qquad (5.61)$$

The Seebeck coefficient is $S = s_T / \sigma$, so

$$S_{\text{tot}} = \frac{S_n \sigma_n + S_p \sigma_p}{\sigma_n + \sigma_p} \,. \qquad (5.62)$$

Since S_n and S_p have opposite signs, we find that for high temperatures, the total S drops, and the performance of a TE device falls.

Exercise 5.2: Evaluation of the 3D, diffusive transport coefficients

We conclude this discussion by working out expressions for the transport coefficients for the 3D, bulk transport equations, eqns. (5.30). The first parameter is the resistivity, $\rho_n = 1/\sigma_n$. According to eqn. (5.27)

$$\sigma_n = \frac{2q^2}{h} \int_{E_c}^{\infty} M_{3D}^c(E) \lambda_n(E) \left(-\frac{\partial f_0}{\partial E} \right) dE \,, \qquad (5.63)$$

which we can rewrite as

$$\sigma_n = \frac{2q^2}{h} \left[\frac{\int_{E_c}^{\infty} M_{3D}^c(E) \lambda_n(E) \left(-\frac{\partial f_0}{\partial E} \right) dE}{\int_{E_c}^{\infty} M_{3D}^c(E) \left(-\frac{\partial f_0}{\partial E} \right) dE} \right] \int_{E_c}^{\infty} M_{3D}^c(E) \left(-\frac{\partial f_0}{\partial E} \right) dE \,.$$

$$(5.64)$$

Now we make the following two definitions:

$$\langle M^c_{3D} \rangle \equiv \int_{E_c}^{\infty} M^c_{3D}(E) \left(-\frac{\partial f_0}{\partial E} \right) dE \qquad (5.65)$$

for the average number of channels near the Fermi level, and

$$\langle\langle \lambda_n \rangle\rangle \equiv \left[\frac{\int_{E_c}^{\infty} M^c_{3D}(E) \lambda_n(E) \left(-\frac{\partial f_0}{\partial E} \right) dE}{\int_{E_c}^{\infty} M^c_{3D}(E) \left(-\frac{\partial f_0}{\partial E} \right) dE} \right] = \frac{\langle M^c_{3D}(E) \lambda_n(E) \rangle}{\langle M^c_{3D}(E) \rangle}, \qquad (5.66)$$

which allows us to write the conductivity as

$$\sigma_n = \frac{2q^2}{h} \langle M^c_{3D} \rangle \langle\langle \lambda_n \rangle\rangle . \qquad (5.67)$$

To evaluate the conductivity, we just need to evaluate the average number of conduction channels and the average mean-free-path.

Assuming parabolic energy bands, $M_{3D}(E)$ is given by eqn. (5.54), and eqn. (5.65) becomes

$$\begin{aligned}
\langle M^c_{3D} \rangle &= \int_{E_c}^{\infty} g_v \frac{m^*_n}{2\pi\hbar^2} (E - E_c) \left(-\frac{\partial f_0}{\partial E} \right) dE \\
&= g_v \frac{m^*_n}{2\pi\hbar^2} k_B T_L \int_{E_c}^{\infty} \left(\frac{E - E_c}{k_B T_L} \right) \left(-\frac{\partial f_0}{\partial E} \right) dE \\
&= M_{3D}(k_B T_L) \int_{E_c}^{\infty} \eta \left(+\frac{\partial f_0}{\partial \eta_F} \right) d\eta \\
&= M^c_{3D}(k_B T_L) \mathcal{F}_0(\eta_F) .
\end{aligned} \qquad (5.68)$$

The quantity, $M^c_{3D}(k_B T_L)$ is eqn. (5.54) evaluated at an energy of $E - E_c = k_B T_L$.

Having evaluated $\langle M^c_{3D} \rangle$, we now turn to $\langle\langle \lambda \rangle\rangle$. Often, we can write the energy dependent mean-free-path as

$$\lambda_n(E) = \lambda_0 [(E - E_c)/k_B T_L]^r , \qquad (5.69)$$

where r is a characteristic exponent that depends on the specific scattering mechanism. We will see in Lecture 6 that this *power law scattering* is reasonable for some common scattering mechanisms.

Using this expression for $\lambda(E)$, we find that the numerator of eqn. (5.66) is

$$
\begin{aligned}
\text{num} &= \int_{E_c}^{\infty} \left[g_v \frac{m_n^*}{2\pi\hbar^2}(E - E_c) \right] \left[\lambda_0 \left(\frac{E - E_c}{k_B T_L} \right)^r \right] \left(-\frac{\partial f_0}{\partial E} \right) dE \\
&= M_{3D}^c (k_B T_L)\lambda_0 \int_{E_c}^{\infty} \left(\frac{E - E_c}{k_B T_L} \right)^{r+1} \left(-\frac{\partial f_0}{\partial E} \right) dE \\
&= M_{3D}^c (k_B T_L) \, \lambda_0 \, \Gamma(r + 2) \, \mathcal{F}_r(\eta_F) \, .
\end{aligned}
\tag{5.70}
$$

Since the denominator of eqn. (5.66) is just $\langle M_{3D} \rangle$, we find

$$
\langle\langle \lambda_n \rangle\rangle = \lambda_0 \, \Gamma(r + 2) \frac{\mathcal{F}_r(\eta_F)}{\mathcal{F}_0(\eta_F)} \, ,
\tag{5.71}
$$

and the conductivity is

$$
\begin{aligned}
\sigma_n &= \frac{2q^2}{h} \langle M_{3D}^c \rangle \langle\langle \lambda_n \rangle\rangle \\
&= \frac{2q^2}{h} M_{3D}^c (k_B T_L) \, \mathcal{F}_0(\eta_F) \left[\lambda_0 \, \Gamma(r + 2) \frac{\mathcal{F}_r(\eta_F)}{\mathcal{F}_0(\eta_F)} \right] \, .
\end{aligned}
\tag{5.72}
$$

Having computed the conductivity (and resistivity) in eqns. (5.30), let's compute the Seebeck coefficient, S_n, next. From eqns. (5.27), we see that

$$
S_n = \left(-\frac{k_B}{q} \right) \left(\frac{E_c - E_F}{k_B T_L} + \delta_n \right) \, ,
\tag{5.73}
$$

where

$$
\delta_n = \frac{\int \left(\frac{E - E_c}{k_B T_L} \right) \sigma_n'(E) dE}{\int \sigma_n'(E) dE} \, .
\tag{5.74}
$$

It is now be straightforward to evaluate this integral and find

$$
\delta_n = (r + 2) \frac{\mathcal{F}_{r+1}(\eta_F)}{\mathcal{F}_r(\eta_F)} \, .
\tag{5.75}
$$

Consider the nondegenerate case where $\eta_F \ll 0$ and both Fermi-Dirac integrals become e^{η_F}. In this case, $\delta_n = (r + 2)$. For $r = 0$, $\delta_n = 2$, and we see that the average energy of current flow is $2k_B T_L$ above E_c. For $r > 0$, the mean-free-path increases with energy, which causes the current to flow at a higher average energy with a corresponding increase in S_n. For $r = 2$, which is the characteristic of ionized impurity scattering, $\delta_n = 4$, and we see that the average energy of current flow is $4k_B T_L$ above E_c.

The final transport coefficient is the electronic thermal conductivity. We leave it as an exercise to show from eqns. (5.27) and (5.31) that

$$L = \frac{\kappa_n}{T_L \sigma_n}$$

$$= \left(\frac{k_B}{q}\right)^2 \left[(r+2)(r+3)\frac{\mathcal{F}_{r+2}(\eta_F)}{\mathcal{F}_r(\eta_F)} - \left((r+2)\frac{\mathcal{F}_{r+1}(\eta_F)}{\mathcal{F}_r(\eta_F)}\right)^2\right]. \tag{5.76}$$

This result should be compared to eqn. (5.53), the general result that does not assume parabolic energy bands or power law scattering. Note that in the non-degenerate limit for $r = 0$, the term in the square brackets is just 2, as asserted earlier. In the degenerate limit, we need to expand the Fermi-Dirac integrals to show that the factor is $\pi^2/3$.

Seebeck coefficient in the degenerate limit: Mott Formula

Equations (5.73)–(5.75) give the Seebeck coefficient for parabolic bands with an arbitrary level of carrier degeneracy. For the non-degenerate limit, the Fermi-Dirac intrgrals become exponentials, and eqn. (5.73) is simplified to

$$S_n = \left(-\frac{k_B}{q}\right)[(r+2) - \eta_F]. \tag{5.77}$$

We can also simplify eqn. (5.73) for strong carrier degeneracy as follows.

In the strongly degenerate limit, the Fermi-Dirac integral of order r approaches (see eqn. (3) in [2]):

$$\mathcal{F}_r(\eta_F) \to \frac{\eta_F^{r+1}}{\Gamma(r+2)} + \frac{\eta_F^{r-1}}{\Gamma(r)}\zeta(2) + \dots, \tag{5.78}$$

where $\Gamma(\bullet)$ is the Gamma function defined in eqn. (3.22), and $\zeta(2) = \pi^2/6$ is the Riemann Zeta function. This expansion can be used in eqn. (5.73),

$$S_n = \left(-\frac{k_B}{q}\right)\left((r+2)\frac{\mathcal{F}_{r+1}(\eta_F)}{\mathcal{F}_r(\eta_F)} - \eta_F\right), \tag{5.79}$$

to find

$$S_n = \left(-\frac{k_B}{q}\right)\left(\frac{2\zeta(2)(r+1)\eta_F}{\eta_F^2 + \zeta(2)(r+1)r}\right), \tag{5.80}$$

which can be simplified for strong degeneracy ($\eta_F \gg 0$) to

$$S_n = \left(-\frac{k_B}{q}\right)\left(\frac{2\zeta(2)(r+1)}{\eta_F}\right). \tag{5.81}$$

We see from eqn. (5.77) that for non-degenerate semiconductors, $|S_n| \propto -\eta_F$ and from eqn. (5.81) that for degenerate semiconductors, $|S_n| \propto 1/\eta_F$. Finally, recall that

$$\sigma'(E) \propto \lambda(E)M_{3D}(E) \propto (E - E_c)^{r+1}, \qquad (5.82)$$

from which we obtain

$$\frac{1}{\sigma'(E)} \frac{d\sigma'(E)}{dE}\Big|_{E=E_F} = \frac{(r+1)}{E_F - E_c}. \qquad (5.83)$$

By making use of eqn. (5.83) in eqn. (5.81), we finally obtain the well-known *Mott relation*

$$\boxed{S_n = -\frac{\pi^2 k_B^2 T_L}{3q} \left(\frac{\partial \ln \sigma'(E)}{\partial E} \right)_{E=E_F}.} \qquad (5.84)$$

The Mott relation is widely-used to describe the Seebeck coefficient in degenerate semiconductors. Note that our starting point for the derivation was eqn. (5.79), which assumes parabolic energy bands and power law scattering, but the same final result can be obtained more generally with the use of the well-known Sommerfeld expansion [2].

5.6 Summary

Our goal in this lecture has been to formally derive the results presented in Lecture 4. In doing so, we have developed general expressions that are valid in 1D, 2D, and 3D for arbitrary band structures and scattering mechanisms and from the ballistic to diffusive limits. We showed how to evaluate these general expressions for two specific cases, 1D ballistic transport and 2D diffusive transport. We also discussed how to treat p-type semiconductors, bipolar conduction, and the Wiedemann-Franz Law, but so far, we have described scattering processes phenomenologically — asserting that $T(E) = \lambda(E)/(\lambda(E) + L)$ and describing $\lambda(E)$ in power law form. In the next lecture, we will discuss scattering in more detail.

5.7 References

Carbon nanotubes can often be treated as ideal, 1D conductors. For an introduction to carbon nanotubes, see:

[1] Mark Lundstrom and Jing Guo, *Nanoscale Transistors: Physics, Modeling, and Simulation*, Springer, New York, 2006.

The essentials of Fermi-Dirac integrals are discussed by Kim, who also discusses the widely-used Sommerfeld expansion, which is used to evaluate transport integrals under strongly degenerate conditions.

[2] R. Kim and M.S. Lundstrom, "Notes on Fermi-Dirac Integrals", 3rd Ed., https://www.nanohub.org/resources/5475.

Lecture 6

An Introduction to Scattering

Contents

6.1 Introduction

We have seen in previous lectures that the mean-free-path for backscattering, λ, plays a key role in near-equilibrium transport. We have asserted that

$$T(E) = \frac{\lambda(E)}{\lambda(E) + L}. \tag{6.1}$$

Where does this expression come from?

We expect that the mean-free-path is the "average" distance that a carrier scatters. In the Landauer approach, the mean-free-path has a specific meaning; it is the length at which the transmission drops to one-half. We shall see that this "mean-free-path for backscattering" is proportional to the actual mean-free-path, Λ,

$$\lambda(E) \propto \Lambda(E) = \upsilon(E)\,\tau(E). \tag{6.2}$$

The material's band structure determines $\upsilon(E)$, and the scattering processes (and band structure) determine the scattering time. Although we will not go into the detailed calculations, we aim to understand how band structure and scattering physics determine λ. We will also discuss how to estimate the average mean-free-path from conductivity or mobility measurements. We begin with a discussion of what controls the scattering time, $\tau(E)$.

6.2   Physics of carrier scattering

Figure 6.1 is a sketch that illustrates some key concepts. An ensemble of carriers is injected into a semiconductor at time, $t = 0$ with energy, E. The initial momenta are aligned along one direction. Some time later, at $t = \tau(E)$, carriers have, on average, experienced a scattering event. Depending on the nature of the scattering event, a carrier's momentum (i.e. the direction of the arrow) may change, and the energy (represented by the length of the arrow) may also increase or decrease. If the scattering mechanism is anisotropic and tends to deflect a carrier by a small angle, then one scattering event will not be enough to eliminate the initial, directed momentum. If we wait longer, however, at time $t = \tau_m(E) \geq \tau(E)$, the momentum will have been randomized. If, however, the dominant scattering mechanism is elastic, then the initial energy of the injected carriers will not have relaxed. Waiting even longer to time $t = \tau_E(E) \gg \tau_m(E), \tau(E)$, we find that the initial excess energy has relaxed. Figure 6.1 illustrates the three characteristic times for scattering: 1) the time between scattering events, $\tau(E)$, 2) the *momentum relaxation time*, $\tau_m(E)$, 3) the *energy relaxation time*, $\tau_E(E)$. In general, $\tau_m(E) \geq \tau(E)$ and $\tau_E(E) \gg \tau(E), \tau_m(E)$.

Since we are interested in the flow of charge and heat currents, we are most interested in the momentum relaxation time and how it depends on the physics of scattering.

The fundamental quantity in a scattering calculation is the transition rate, $S(\vec{p} \rightarrow \vec{p}\,')$, from an initial state, $\vec{p}$, to one specific final state, $\vec{p}\,'$.

The total scattering rate, the probability per unit time of scattering, is just one over the average time between collisions and is obtained by summing over all of the possible final states, $\vec{p}\,'$ that carriers may scatter to. The result is

$$\frac{1}{\tau(\vec{p})} = \sum_{\vec{p}\,'} S(\vec{p} \rightarrow \vec{p}\,') \,. \tag{6.3}$$

Similarly, to get the momentum relaxation time, we weight by the fractional change in momentum for each scattering event. Assuming that the initial

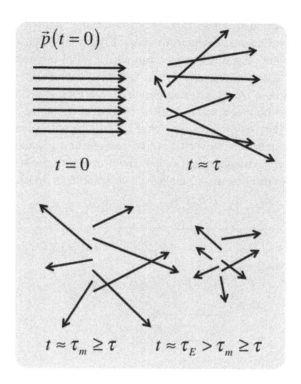

$\vec{p}(t=0)$

$t = 0$

$t \approx \tau$

$t \approx \tau_m \geq \tau$

$t \approx \tau_E > \tau_m \geq \tau$

Fig. 6.1. Sketch illustrating the characteristic times for carrier scattering. An ensemble of carriers with momentum directed along one axis is injected at $t = 0$. Carriers have, on average, experienced one collision at $t = \tau(E)$. The momentum of the initial ensemble has been relaxed to zero at $t = \tau_m(E)$, and the energy has relaxed to its equilibrium value at $t = \tau_E(E)$. (After Lundstrom, [1]).

momentum is directed along the z-axis, we find

$$\frac{1}{\tau_m(\vec{p})} = \sum_{\vec{p}'} S(\vec{p} \rightarrow \vec{p}') \frac{\Delta p_z}{p_z}. \tag{6.4}$$

Anisotropic scattering tends to deflect carriers by small angles, which produces a small fractional change in the incident momentum, reducing the momentum relaxation rate and increasing the momentum relaxation time. The energy relaxation rate would be given by a similar expression, but with the fractional change in momentum replaced by the fractional change in energy.

Since we can calculate the momentum relaxation time from the transition rate, we just need to understand how the transition rate is calculated.

For an introduction to these types of calculations, see [1]. Here we will just summarize how the calculation goes. Figure 6.2 is an illustration of a scattering event. An incident electron with crystal momentum, $\vec{p} = \hbar\vec{k}$, described by a Bloch wavefunction, $\psi_i(\vec{r})$, enters a region over which the potential is perturbed by a scattering potential, $U_S(\vec{r}, t)$, which may be static (as for charged impurity scattering) or time dependent (as for scattering from phonons (lattice vibrations)). After interacting with the scattering potential, the electron emerges with a final momentum, $\vec{p}'$, and Bloch state, $\psi_f(\vec{r})$. We need to calculate $S(\vec{p} \to \vec{p}')$, the probability per second that the electron in the initial state, $\vec{p}$, makes a transition to a specific final state, $\vec{p}'$.

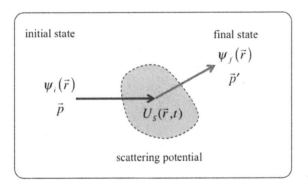

Fig. 6.2. Illustration of a scattering event. An initial electron in state, $\vec{p}$, with wavefunction, ψ_i, interacts with a scattering potential, $U_S(\vec{r}, t)$ and emerges in the state, $\vec{p}'$, described by the wavefunction, ψ_f.

The most common way to calculate transition rates for carrier scattering in semiconductors is by first order perturbation theory — Fermi's Golden Rule (FGR) of quantum mechanics.

The prescription is

$$S(\vec{p} \to \vec{p}') = \frac{2\pi}{\hbar} |H_{p',p}|^2 \, \delta \left(E' - E - \Delta E \right) , \tag{6.5}$$

where the matrix element is

$$|H_{p',p}|^2 = \int_{-\infty}^{+\infty} \psi_f^*(\vec{r}) \, U_S(\vec{r}) \, \psi_i(\vec{r}) d\vec{r} . \tag{6.6}$$

Note the order of the subscripts in the matrix element, final state first, then initial state. Section 1.7 of Ref. [1] gives a derivation of Fermi's Golden

Rule and discusses the approximation involved [1]. The matrix element in eqn. (6.5) couples the strength of the scattering potential to the deflection of a carrier. The δ-function is a statement of energy conservation. For a static scattering potential, such as charged impurity scattering, $\Delta E = 0$, and there is no energy relaxation. For a scattering potential that is vibrating at a frequency of ω, such as a lattice vibration, $\Delta E = \pm \hbar \omega$ depending on whether a phonon is absorbed or emitted.

Scattering rate calculations proceed as follows. First, the scattering potential must be identified, i.e. $U_S(\vec{r}, t)$ must be identified for the specific scattering process in question. Then the transition rate can be evaluated from FGR as given by eqn. (6.5). Next, the characteristic times are evaluated according to eqns. (6.3) and (6.4). Later on, we will see how the mean-free-path for backscattering can be obtained from the momentum relaxation time. To see how such calculations are done, refer to Chapter 2 in [1]. Without doing the actual calculations, however, it is useful to get a general feel for the results.

Some simple scattering potentials (e.g. short range scattering that can be described by a δ-function scattering potential, and acoustic and optical phonon scattering in nonpolar materials) simply deflect with equal probability of the incident carriers to final states that conserve energy. In such cases, we expect that the scattering rate will be proportional to the density of final states. For elastic scattering, we have $1/\tau(E) \propto D(E)$, for scattering by phonon absorption, $1/\tau(E) \propto D(E + \hbar \omega)$, and for scattering by phonon emission, $1/\tau(E) \propto D(E - \hbar \omega)$. Since the density of states generally increases with energy, we expect the scattering time to decrease with increasing energy of the incident carrier.

For scattering from charged impurities or from phonons in polar materials, it is different. As illustrated in Fig. 6.3, randomly located charges introduce fluctuations into the bottom of the conduction band, $E_c(\vec{r})$, which can scatter carriers. High energy carriers, however, do not feel this fluctuating potential as much as low energy carriers, so for charged impurity (and polar phonon) scattering, we expect that $1/\tau(E)$ will decrease (the scattering time, $\tau(E)$ will increase) as the carrier energy increases. For nonpolar phonon scattering, the scattering time decreases with energy, but for charged impurity and polar phonon scattering, it increases with energy.

For some common scattering mechanisms, the scattering time can be written as (or approximately written as)

$$\tau(E) = \tau_0 \left(\frac{E - E_c}{k_B T_L} \right)^s, \tag{6.7}$$

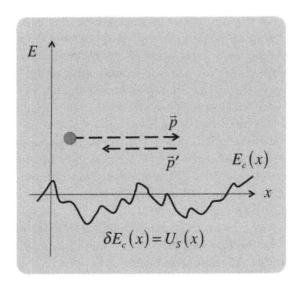

Fig. 6.3.   Illustration of charged impurity scattering. High energy carriers feel the perturbed potential less than low energy carriers and are, therefore, scattered less.

where s is a characteristic exponent that describes the specific scattering mechanism. Equation (6.7) is known as *power law scattering* and is commonly used to find analytical solutions to problems. For acoustic phonon scattering in 3D with parabolic bands, $s = -1/2$, and for ionized impurity scattering, $s = +3/2$ [1].

Our goal in this lecture is to understand how scattering affects the mean-free-path and transmission. We expect that the mean-free-path will be proportional to the product of velocity and scattering time. We now have a general understanding of what determines the scattering time. Before we relate the mean-free-path to scattering time, let's first relate the mean-free-path to the transmission.

6.3   Transmission and mean-free-path

To relate the mean-free-path to transmission, consider the simple problem sketched in Fig. 6.4. We assume a slab with a uniform mean-free-path, λ. No electric field is present, so $E_c(x)$ is constant. A flux of electrons, $I^+(x = 0)$, is injected from the left. Some fraction, T, emerges from the right, $I^+(x = L) = TI^+(x = 0)$. The rest of the injected flux, $I^-(x =$

$0) = RI^+(x = 0)$, is backscattered and emerges from the left. If none of the injected flux recombines within the slab, then $T + R = 1$. The contact at the right is an absorbing contact — flux can emerge from the slab, but no flux is injected from the right contact. The net current, therefore, is $I = (1 - R)I^+(0) = TI^+(0)$. Within the slab, we have both positive and negative-directed fluxes, and we seek first to describe their spatial dependence.

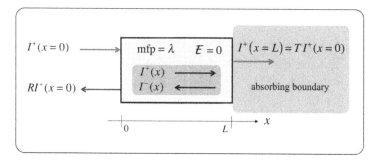

Fig. 6.4.   A model calculation for transmission. A flux is injected at the left of a slab having a mean-free-path for backscattering of λ, and we seek to compute the flux that emerges from the right.

We now define $1/\lambda$ to be the probability per unit length that a positive flux is converted into a negative flux (or vice versa). This is why we call λ the mean-free-path for backscattering. Within the slab, some of the injected positive flux is converted to a negative flux by backscattering. A negatively-directed flux builds up within the slab, and some of the negative flux backscatters and increases the positively-directed flux. Accordingly, the mathematical description for the positively-directed flux within the slab is

$$\frac{dI^+(x)}{dx} = -\frac{I^+(x)}{\lambda} + \frac{I^-(x)}{\lambda}. \tag{6.8}$$

Assuming that there is no recombination or generation within slab, we also see that $I = I^+(x) - I^-(x)$ is a constant, which can be used to rewrite eqn. (6.8) as

$$\frac{dI^+(x)}{dx} = -\frac{I}{\lambda}. \tag{6.9}$$

Equation (6.9) shows that the positively-directed flux decays linearly within the slab.

Now eqn. (6.9) can be integrated from $x = 0$ to some point within the slab to find

$$I^+(x) = I^+(0) - I\frac{x}{\lambda},\tag{6.10}$$

which can be used to find the flux emerging from the right as

$$\begin{aligned}I^+(L) &= I^+(0) - I\frac{L}{\lambda}\\ &= I^+(0) - \left(I^+(L) - I^-(L)\right)\frac{L}{\lambda}\\ &= I^+(0) - I^+(L)\frac{L}{\lambda}\,.\end{aligned}\tag{6.11}$$

In the last line, we made use of the fact that no flux is injected from the right side of the slab, $I^-(L) = 0$. Finally, we can solve eqn. (6.11) for

$$I^+(L) = \frac{\lambda}{\lambda + L}I^+(0) = TI^+(0)\,.\tag{6.12}$$

If we repeat the calculation for the opposite case where we inject a flux, $I^-(L)$ from the right, we find that the flux emerging from the left is $I^-(0) = T'I^-(L)$. Since we have assumed a uniform slab, $T' = T$. A slab under bias is not uniform, but near-equilibrium, we can assume that $T' \approx T$. Finally, if we resolve the incident flux in energy, and assume that energy channels are nearly independent, then we arrive at our final result,

$$\boxed{T(E) = \frac{\lambda(E)}{\lambda(E) + L}\,.}\tag{6.13}$$

Although this is a simple calculation with several simplifying approximations, the author's experience is that it typically works very well in practice. Our derivation has also taught us that the interpretation of the mean-free-path is that its inverse is the probability per unit length that a positive flux is converted into a negative flux. That is why we call it a mean-free-path for backscattering.

6.4   Mean-free-path and scattering

We are now ready to relate the mean-free-path for backscattering to the scattering time. The distinction between mean-free-path and mean-free-path for backscattering is easiest to see in 1D. Consider Fig. 6.5 which shows a 1D scattering event. An incident electron undergoes a scattering event. If the scattering is isotropic, there are two possibilities, the electron can forward scatter or back scatter. Only backscattering is relevant for the mean-free-path for scattering, so the time between backscattering events is $2\tau_m$. We conclude that the mean-free-path for backscattering is twice the mean-free-path for scattering,

$$\lambda(E) = 2\Lambda = 2v(E)\tau_m(E). \tag{6.14}$$

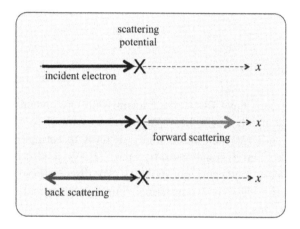

Fig. 6.5. Illustration of forward and backscattering in 1D.

It is a little harder to see the relation of the two mean-free-paths in 2D and 3D, but it can be shown that the proper definition of the mean-free-path for backscattering is [2]

$$\lambda(E) = 2\frac{\langle v_x^2 \tau_m \rangle}{\langle |v_x| \rangle}, \tag{6.15}$$

where the brackets denote an average over angle at the energy, E. Working this expression out for isotropic bands, we find

$$\boxed{\begin{array}{ll} \lambda(E) = 2v(E)\tau_m(E) & \text{1D} \\ \lambda(E) = \dfrac{\pi}{2}v(E)\tau_m(E) & \text{2D} \\ \lambda(E) = \dfrac{4}{3}v(E)\tau_m(E) & \text{3D}. \end{array}} \tag{6.16}$$

In Lecture 7, we will see where these factors come from.

Finally, it is frequently useful to write the mean-free-path in power law form as

$$\lambda(E) = \lambda_0 \left(\frac{E - E_c}{k_B T_L} \right)^r , \qquad (6.17)$$

where r is a characteristic exponent. We used a similar expression to describe the scattering time in eqn. (6.7) with a characteristic exponent, s. For parabolic energy bands, $v(E) \propto E^{1/2}$, so $r = s + 1/2$. Consequently, $r = 0$ for acoustic phonon scattering, and $r = 2$ for ionized impurity scattering.

6.5   Discussion

In this lecture, we have discussed transmission, the mean-free-path for backscattering, and how the mean-free-path is related to the physics of scattering. In practice, one often analyzes data to estimate the average mean-free-path. The average mean-free-path, $\langle\langle\lambda\rangle\rangle$, is also related to the measured diffusion coefficient and mobility. We discuss these topics in this section. Only the 2D case is considered, but similar considerations apply in other dimensions.

Estimating average mfp from measurements

For a 2D, diffusive conductor,

$$G_{2D} = \sigma_S \frac{W}{L} , \qquad (6.18)$$

where the sheet conductance is given by

$$\sigma_S = \frac{2q^2}{h} \int M_{2D}(E) \lambda(E) \left(-\frac{\partial f_0}{\partial E} \right) dE . \qquad (6.19)$$

Assuming that we have measured σ_S, how can we estimate $\langle\langle\lambda\rangle\rangle$?

Equation (6.19) can be expressed as

$$\sigma_S = \frac{2q^2}{h} \langle M_{2D} \rangle \langle\langle\lambda\rangle\rangle , \qquad (6.20)$$

where $\langle M_{2D} \rangle$ is given by eqn. (3.18) and $\langle\langle\lambda\rangle\rangle$ by eqn. (3.30). Having measured σ_S, we seek to determine $\langle\langle\lambda\rangle\rangle$ from

$$\langle\langle\lambda\rangle\rangle = \frac{\sigma_S}{(2q^2/h) \langle M_{2D} \rangle} . \qquad (6.21)$$

According to eqn. (3.18),

$$\langle M_{2D} \rangle = \left(\frac{\sqrt{\pi}}{2} \right) M_{2D}(k_B T_L) \mathcal{F}_{-1/2}(\eta_F), \qquad (6.22)$$

but we do not know the location of the Fermi level, or η_F. Just measuring the conductivity is not enough, we must also measure the sheet carrier density,

$$n_s = \left(g_v \frac{m^* k_B T_L}{\pi \hbar^2} \right) \mathcal{F}_0(\eta_F) = N_{2D} \mathcal{F}_0(\eta_F). \qquad (6.23)$$

From the measured sheet carrier density, we can deduce η_F and insert the result in eqn. (6.22) to find $\langle\langle \lambda \rangle\rangle$ from eqn. (6.21). For a non-degenerate semiconductor, the Fermi-Dirac integrals reduce to exponentials, and we can get an explicit expression for $\langle\langle \lambda \rangle\rangle$ in terms of the measured sheet conductance and sheet carrier density,

$$\langle\langle \lambda \rangle\rangle = \frac{2 (k_B T_L/q)}{q v_T} \left(\frac{\sigma_S}{n_S} \right), \qquad (6.24)$$

where v_T is the unidirectional thermal velocity as given by eqn. (3.69). If we can measure the sheet conductance and the sheet carrier density, then we can experimentally determine the average mean-free-path for backscattering. Frequently, however, the experimental results are given in terms of the measured diffusion coefficient or mobility, so we should relate $\langle\langle \lambda \rangle\rangle$ to those quantities.

Relating the mfp to the diffusion coefficient

Consider again our model problem for computing the transmission, as sketched in Fig. 6.4. We may think of this as a diffusion problem — a flux of carriers is injected at the left and collected at the right. What is the diffusion coefficient for this problem, and how does it relate to the mean-free-path?

Recall that a flux in the x direction is the product of the density of particles and their average velocity in the direction of transport. At $x = 0$, the number of electrons moving in the $+x$ direction is $n^+(0) = I^+(0)/\langle v_x^+ \rangle$, where $\langle v_x^+ \rangle$ is the average velocity of electrons moving in the $+x$ direction. Similarly, $n^-(0) = I^-(0)/\langle v_x^+ \rangle$, where we have assumed near-equilibrium conditions for which $\langle v_x^+ \rangle \approx \langle v_x^- \rangle$. The total carrier density is just the sum of the two, or

$$n(0) = \frac{(1 + R)\, I^+(0)}{\langle v_x^+ \rangle}$$

$$= \frac{(2 - T)\, I^+(0)}{\langle v_x^+ \rangle}.$$

(6.25)

At the right end of the slab, we have $n^+(L) = I^+(L)/\langle v_x^+ \rangle$, and $n^-(L) = 0$, because no electrons are injected from the right contact. The total carrier density at $x = L$ is

$$n(L) = \frac{I^+(L)}{\langle v_x^+ \rangle} = \frac{T I^+(0)}{\langle v_x^+ \rangle},$$

(6.26)

which is less than $n(0)$. From eqns. (6.25) and (6.26), we find

$$n(0) - n(L) = \frac{I^+(0)}{\langle v_x^+ \rangle} 2(1 - T).$$

(6.27)

Using the fact that the total current is $I = T I^+(0)$, we can solve eqn. (6.27) to find the current as

$$I = \frac{\langle v_x^+ \rangle}{2} \frac{TL}{1 - T} \times \left[\frac{n(0) - n(L)}{L} \right] = -\frac{\langle v_x^+ \rangle \lambda}{2} \times \frac{dn(x)}{dx}.$$

(6.28)

Finally, we define the diffusion coefficient, D_n, and recognize the result as the well-known Fick's Law of diffusion:

$$\boxed{\begin{aligned} I &= -D_n \frac{dn}{dx} \\ D_n &= \frac{\langle v_x^+ \rangle \lambda}{2}. \end{aligned}}$$

(6.29)

Our final result is the familiar Fick's Law of diffusion, but it is actually a surprising result. It is often said that Fick's Law only holds for diffusion across regions that are many mean-free-paths long, but we made no such restriction in our derivation. William Shockley pointed in 1962 that Fick's Law is not restricted to long regions [3]; it can be used to described ballistic and quasi-ballistic "diffusion"; we just need to be careful about the boundary conditions.

More generally, we can view our derivation as being for electrons in an energy channel, E. To get the total current, we would integrate over the energy channels, and that would relate the overall diffusion coefficient to the energy-averaged mean-free-path. As a simpler example, consider a non-degenerate semiconductor with an energy independent mean-free-path, λ_0. The average x-directed velocity is $v_T = \sqrt{2k_BT_L/\pi m^*}$, so eqn. (6.29) becomes

$$\boxed{D_n = \frac{v_T\lambda_0}{2},} \tag{6.30}$$

which provides a simple means to extract the mean-free-path from the measured diffusion coefficient. More general expressions, considering Fermi-Dirac statistics and energy-dependent scattering can be derived.

Relating the mfp to the mobility

The measured mobility is often reported, and it can also be used to estimate the average mean-free-path. Equation (6.19) gives the 2D sheet conductance, which can also be written as $n_S\,q\,\mu_n$. Equating these two expressions, we find

$$\mu_n \equiv \frac{\frac{2q}{h}\int M_{2D}(E)\lambda(E)\left(-\frac{\partial f_0}{\partial E}\right)dE}{n_S}, \tag{6.31}$$

which we take as the definition of mobility. (An equivalent formula is known as the Kubo-Greenwood formula.) We can write eqn. (6.31) as

$$\mu_n = \frac{1}{n_S}\frac{2q}{h}\langle\langle\lambda\rangle\rangle\langle M_{2D}\rangle, \tag{6.32}$$

where $\langle M_{2D}\rangle$ is given by eqn. (6.22). Proceeding as before, we can solve eqn. (6.32) for $\langle\langle\lambda\rangle\rangle$ and find

$$\langle\langle\lambda\rangle\rangle = \frac{2\,(k_BT_L/q)\,\mu_n}{v_T}\times\frac{\mathcal{F}_0(\eta_F)}{\mathcal{F}_{-1/2}(\eta_F)}. \tag{6.33}$$

Given a measured mobility and carrier density (needed to find η_F), the average mean-free-path can be estimated. Under non-degenerate conditions, the last factor is one and we find

$$\langle\langle\lambda\rangle\rangle = \frac{2\,(k_BT_L/q)\,\mu_n}{v_T}, \tag{6.34}$$

so for a non-degenerate semiconductor, it is easy to determine $\langle\langle\lambda\rangle\rangle$.

Solving eqn. (6.34) for the mobility of a non-degenerate semiconductor, we find

$$\mu_n = \frac{v_T\,\langle\langle\lambda\rangle\rangle}{2}\times\frac{1}{(k_BT_L/q)}. \tag{6.35}$$

Now by defining the diffusion coefficient as

$$D_n = \frac{\upsilon_T \langle\langle\lambda\rangle\rangle}{2} .$$ (6.36)

we see that

$$\boxed{\frac{D_n}{\mu_n} = \frac{k_B T_L}{q} ,}$$ (6.37)

which is just a statement of the Einstein relation for non-degenerate conditions.

Average mean-free-path for power law scattering

According to eqns. (3.30) and (3.31), in 2D, the average mean-free-path for power law scattering is given by

$$\langle\langle\lambda\rangle\rangle = \lambda_0 \frac{\int \left(\frac{E-E_c}{k_B T_L}\right)^r M_{2D}(E) \left(-\frac{\partial f_0}{\partial E}\right) dE}{\int M_{2D}(E) \left(-\frac{\partial f_0}{\partial E}\right) dE} ,$$ (6.38)

where $M_{2D}(E)$ is given by eqn. (3.18). This integral is readily worked out to find

$$\langle\langle\lambda\rangle\rangle = \lambda_0 \frac{\Gamma(r + 3/2)}{\Gamma(3/2)} \times \frac{\mathcal{F}_{r-1/2}(\eta_F)}{\mathcal{F}_{1/2}(\eta_F)} .$$ (6.39)

For a non-degenerate semiconductor, the last factor is one.

Exercise 6.1: Mobility for a constant scattering time

As a quick example of using this result, consider the mobility for a non-degenerate semiconductor with a constant scattering time, τ_0. From eqn. (6.35), we have

$$\mu_n = \frac{\upsilon_T \langle\langle\lambda\rangle\rangle}{2} \frac{1}{(k_B T_L/q)} = \frac{\upsilon_T \lambda_0 \left(\frac{\Gamma(r+3/2)}{\Gamma(3/2)}\right)}{2} \frac{1}{(k_B T_L/q)} .$$ (6.40)

From eqn. (6.16), we find

$$\lambda(E) = \frac{\pi}{2}\upsilon(E)\tau_0 = \left[\frac{\pi}{2}\sqrt{\frac{2k_B T_L}{m^*}}\right] \tau_0 \left(\frac{E - E_c}{k_B T_L}\right)^{1/2} ,$$ (6.41)

from which, we see that $r = 1/2$ and $\lambda_0 = \pi \sqrt{2k_B T_L/m^*}\, \tau_0/2$. Using these results in eqn. (6.40), we find

$$\boxed{\mu_n = \frac{q\,\tau_0}{m^*},}$$
(6.42)

which is the expected result.

Exercise 6.2: Estimating the mean-free-path for a MOSFET

In Exercise 3.1 we considered a nanoscale MOSFET, and in Exercise 3.2, we estimated the average mean-free-path of about 40 nm. This analysis made use of measured channel resistance at 300 K and used a $T_L = 0$ K expression to simplify the analysis. Let's see if we can do a better job of estimating the room temperature $\langle\langle \lambda \rangle\rangle$ from the reported room temperature mobility of 260 cm^2/V-s at a carrier density of $n_S = 6.7 \times 10^{12}$/cm^2.

Let's first do the calculation with Maxwell-Boltzmann (non-degenerate) carrier statistics. Using the Einstein relation, we find the diffusion coefficient as

$$D_n = \frac{k_B T_L}{q}\mu_n = 6.7 \text{ cm}^2/\text{s}.$$
(6.43)

Now we can use eqn. (6.36) to determine $\langle\langle \lambda \rangle\rangle$. To evaluate v_T, we need the effective mass. For electrons in the inversion layer of (100) Si, when only one subband of conduction band occupied, the effective mass is the transverse mass, $m^* = 0.19\, m_0$ [4] from which we find $v_T = 1.2 \times 10^7$ cm/s. Using this value in eqn. (6.36), we find

$$\langle\langle \lambda \rangle\rangle_{MB} \approx 11 \text{ nm}.$$
(6.44)

Now let's redo the calculation with Fermi-Dirac statistics. From eqn. (6.33) we find

$$\langle\langle \lambda \rangle\rangle = \langle\langle \lambda \rangle\rangle_{MB} \times \frac{\mathcal{F}_0(\eta_F)}{\mathcal{F}_{-1/2}(\eta_F)}.$$
(6.45)

To find η_F, recall that the sheet carrier density is given by

$$n_S = N_{2D}\mathcal{F}_0(\eta_F) = \left(g_v \frac{m^* k_B T_L}{\pi \hbar^2} \right) \mathcal{F}_0(\eta_F).$$
(6.46)

For electrons in the first subband of our (100) oriented Si inversion layer, $g_v = 2$, so we find $N_{2D} = 4.1 \times 10^{11}$/cm^2. Recall that the Fermi-Dirac integral of order 0 is analytical, $\mathcal{F}_0(\eta_F) = \ln(1 + e^{\eta_F})$, so we can solve for η_F and find

$$\eta_F = \ln(e^{n_S/N_{2D}} - 1) = 1.42. \tag{6.47}$$

Finally, we find that

$$\langle\langle\lambda\rangle\rangle = 11\,\text{nm} \times \frac{\mathcal{F}_0(1.42)}{\mathcal{F}_{-1/2}(1.42)} \approx 15\,\text{nm}, \tag{6.48}$$

which is our best estimate of $\langle\langle\lambda\rangle\rangle$ for this transistor in the on-state. Note that the channel length of the transistor was 60 nm, which is neither long nor short compared to the mean-free-path. This transistor operates in the quasi-ballistic regime.

6.6 Summary

Our goal in this lecture has been to discuss how transmission is related to the mean-free-path for backscattering and how the mean-free-path is related to carrier scattering. We also discussed how the average mean-free-path can be estimated from measurements of the conductivity (or mobility) and the carrier density. The discussion has been restricted to 2D and parabolic energy bands, but similar considerations apply to other dimensions and band structures.

This lecture completes our discussion of the Landauer theory of low field transport. In the next lecture, we will discuss an older and still more common approach.

6.7 References

The fundamentals of carrier scattering are discussed in:

[1] Mark Lundstrom, *Fundamentals of Carrier Transport 2nd Ed.*, Cambridge Univ. Press, Cambridge, UK, 2000.

To see where eqn. (6.15), the definition for the mean-free-path for backscattering, comes from, see

[2] Changwook Jeong, Raseong Kim, Mathieu Luisier, Supriyo Datta, and Mark Lundstrom, "On Landauer vs. Boltzmann and Full Band vs. Effective Mass Evaluation of Thermoelectric Transport Coefficients", *J. Appl. Phys.*, **107**, 023707, 2010.

Our derivation of the diffusion coefficient in terms of the mean-free-path follows an approach due to Shockley.

[3] W. Shockley, "Diffusion and Drift of Minority Carriers in Semiconductors for Comparable Capture and Scattering Mean Free Paths", *Phys. Rev.*, **125**, 1570-1576, 1962.

For a discussion of the electron subband structure, valley degeneracy, and effective masses for (100) Si inversion layers, see

[4] Yuan Taur and Tak Ning, *Fundamentals of Modern VLSI Devices 2nd Ed.*, Cambridge Univ. Press, Cambridge, UK, 2009.

Boltzmann Transport Equation

Contents

7.1   Introduction

In this lecture we discuss a different approach to carrier transport, the Boltzmann Transport Equation (BTE). For some problems, the BTE is the better choice and for some, the Landauer approach is preferred. Of course, when we solve the same problem, we get the same answers. This lecture is an introduction to the BTE and its relation to the Landauer approach. As an application of the BTE, we discuss how magnetic fields affect carrier transport.

Our goal is to find the *distribution function*, $f(\vec{r}, \vec{k}, t)$, the probability that a state at position, $\vec{r}$, with wavevector, $\vec{k}$ (or *crystal momentum*, $\vec{p} = \hbar\vec{k}$) is occupied at time, t. The answer in equilibrium is the Fermi function, $f_0(E)$; we seek the solution out of equilibrium. Our use of the wavevector, $\vec{k}$, suggests that we are considering transport in crystalline materials in which the periodic crystal potential leads to Bloch wave solutions. The general model for transport developed in Lecture 2 did not make this assumption, but throughout these lectures we have considered transport in crystalline solids, so this assumption will not be a limitation for us.

Our goals are to:

(1) Find an equation for $f(\vec{r}, \vec{k}, t)$ out of equilibrium.
(2) Learn how to solve the resulting equation (the BTE) under near-equilibrium conditions.
(3) Relate the results to those obtained from the Landauer approach in the diffusive limit.
(4) Add a magnetic field and see how transport changes.

This lecture is a short introduction to the BTE. For a fuller account, see Ashcroft and Mermin, Chapter 13 [1], Ziman [2], or Lundstrom, Chapters 3 and 4 [3].

7.2   The Boltzmann Transport Equation

Electrons in crystals with a slowly varying applied potential can be treated as semi-classical quasi-particles. The equation of motion is like Newton's Law,

$$\frac{d\vec{p}}{dt} = \frac{d(\hbar\vec{k})}{dt} = \vec{F}_e \,, \tag{7.1}$$

where $\hbar\vec{k}$ is the crystal momentum, and $\vec{F}_e$ is the force on the electron,

$$\vec{F}_e = -\nabla E_c - q\vec{v} \times \vec{B} = -q\vec{\mathcal{E}} - q\vec{v} \times \vec{B} \,. \tag{7.2}$$

To find $\vec{k}(t)$ for the electron, we solve

$$\hbar\vec{k}(t) = \hbar\vec{k}(0) + \int_0^t \vec{F}_e\left[\vec{r}(t'), t'\right] dt' \,. \tag{7.3}$$

In addition to tracking electrons in momentum (or $\vec{k}$) space, we must also track their positions. Knowing $\vec{k}(t)$, we determine the electron's velocity from its band structure according to

$$\vec{v}_g(t) = \frac{1}{\hbar}\nabla_k E\left[\vec{k}(t)\right] \,. \tag{7.4}$$

To find $\vec{r}(t)$ for the electron, we solve

$$\vec{r}(t) = \vec{r}(0) + \int_0^t \vec{v}_g(t')dt' \,. \tag{7.5}$$

Equations (7.3) and (7.5) are the semi-classical equations of motion for electrons in a crystalline solid. In the semi-classical approach, we assume

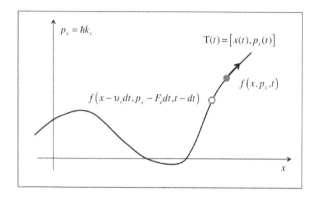

Fig. 7.1. Sketch of an electron trajectory in two-dimentional, $x - p_x$, phase space. We wish to determine the probability that the state indicated by the filled circle is occupied at time, t. This probability is the probability that the corresponding upstream state indicated by the open circle was occupied at time, $t - dt$.

that the potential, $E_c(\vec{r})$, varies slowly, so that there are no quantum mechanical reflections or tunnelling, and we treat electrons as particles. We specify the electron's position and momentum simultaneously and assume that the quantum mechanical uncertainty, $\Delta r \Delta p \geq \hbar/2$, is small.

The equations of motion describe an electron's position vs. time in position-momentum or *phase* space. Figure 7.1 is a sketch of an electron trajectory, $T[x(t), p_x(t)]$, in a two-dimensional phase space. Also shown is a particular state whose occupation probability we wish to determine. The probability that this state is occupied at time, t, is $f(x, p_x, t)$. Since electrons simply move along the trajectory, $T[x(t), p_x(t)]$, the probability that this state is occupied at time, t, is the probability that the corresponding upstream state was occupied at $t = t - dt$, so

$$f(x, p_x, t) = f(x - v_x dt, p_x - F_e dt, t - dt), \tag{7.6}$$

which is simply a statement that the total derivative along the trajectory is zero,

$$\frac{df}{dt} = 0. \tag{7.7}$$

If we use the chain rule to expand eqn. (7.7), we find

$$\frac{df}{dt} = \frac{\partial f}{\partial t} + \frac{\partial f}{\partial x} \frac{\partial x}{\partial t} + \frac{\partial f}{\partial p_x} \frac{\partial p_x}{\partial t} = 0, \tag{7.8}$$

which can be written as

$$\frac{\partial f}{\partial t} + \frac{\partial f}{\partial x}v_x + \frac{\partial f}{\partial p_x}F_x = 0\,. \tag{7.9}$$

In 3D position and 3D momentum phase space, eqn. (7.9) becomes

$$\frac{\partial f}{\partial t} + \vec{v}\cdot\nabla_r f + \vec{F}_e\cdot\nabla_p f = 0\,, \tag{7.10}$$

where $\vec{F}_e$ is given by eqn. (7.2) and

$$\nabla_r f = \frac{\partial f}{\partial x}\hat{x} + \frac{\partial f}{\partial y}\hat{y} + \frac{\partial f}{\partial z}\hat{z} \tag{7.11a}$$

$$\nabla_p f = \frac{\partial f}{\partial p_x}\hat{p}_x + \frac{\partial f}{\partial p_y}\hat{p}_y + \frac{\partial f}{\partial p_z}\hat{p}_z\,. \tag{7.11b}$$

Equation (7.10) is the collisionless Boltzmann Transport Equation; it describes ballistic transport (when there is no scattering) or equilibrium (when each scattering event is cancelled by its inverse according to the principle of detailed balance). Figure 7.2 illustrates the situation when carriers scatter. *In-scattering* from other states increases $f(\vec{r}, \vec{k}, t)$ and *out-scattering* to other states decreases $f(\vec{r}, \vec{k}, t)$. In the semi-classical approach to carrier transport, we assume that scattering is the result of short range forces and compute the scattering rates using Fermi's Golden Rule as discussed in Sec. 6.2. Scattering processes happen quickly, so we assume that the electron's position does not change during a scattering event. Scattering simply changes the momentum of the electron. In the absence of scattering, $df/dt = 0$, but when scattering occurs, we write $df/dt|_{\text{coll}} = \hat{C}f$, where $\hat{C}$ is the *collision operator*. In the presence of scattering, the BTE, eqn. (7.10), becomes

$$\boxed{\frac{\partial f}{\partial t} + \vec{v}\cdot\nabla_r f + \vec{F}_e\cdot\nabla_p f = \hat{C}f\,.} \tag{7.12}$$

The treatment of carrier scattering can get complicated (see [3] for an introduction), so a simple approximation is frequently used. In the *Relaxation Time Approximation* (RTA), we write the collision operator as

$$\hat{C}f = -\left(\frac{f(\vec{p}) - f_0(\vec{p})}{\tau_m}\right) = -\frac{\delta f(\vec{p})}{\tau_m}\,, \tag{7.13}$$

where δf is the deviation of f from its equilibrium value, and τ_m is a characteristic time that turns out to be the momentum relaxation time [3].

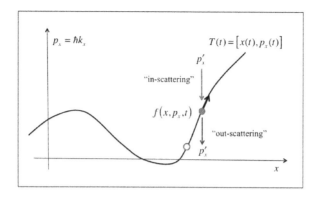

Fig. 7.2. Illustration of how in-scattering and out-scattering affect the occupation probability of a state in phase space.

To get a feel for the RTA, consider a spatially homogeneous material in the absence of an electric or magnetic field. The second and third terms in eqn. (7.12) are zero, so using the RTA, eqn. (7.12) becomes

$$\frac{\partial f}{\partial t} = \frac{\partial(\delta f)}{\partial t} = -\frac{\delta f}{\tau_m}, \tag{7.14}$$

where we used $\partial f_0/\partial t = 0$. Equation (7.14) has the solution,

$$\delta f(t) = \delta f(0)e^{-t/\tau_m}. \tag{7.15}$$

According to eqn. (7.15), perturbations from equilibrium decay away exponentially with a characteristic time, τ_m. Scattering acts to restore equilibrium. The RTA is a reasonable approach to near-equilibrium transport, but this approximation to the collision integral can only be justified near equilibrium and even then only for specific types of scattering (i.e. scattering that is either elastic, isotropic, or both) [3]. A proper treatment of scattering significantly complicates solving the BTE [3]. Although the RTA is not generally applicable, it is widely used to obtain analytical answers and generally produces sensible results. We will adopt the RTA for the remainder of this lecture, but one should remember that it can fail, sometimes in important ways (for an example, see [4]).

7.3 Solving the steady-state BTE

As an example of solving the BTE, let's consider a simple example. We assume steady-state with no magnetic field. Assuming the Relaxation Time

Approximation for the collision integral, the steady-state BTE is

$$\vec{v} \cdot \nabla_r f - q\vec{\mathcal{E}} \cdot \nabla_p f = -\frac{\delta f}{\tau_m}. \tag{7.16}$$

The solution consists of a large part, $f_0(\vec{r}, \vec{p}, t)$, and a small deviation from equilibrium, $\delta f(\vec{r}, \vec{p}, t)$, so it is reasonable to assume

$$\begin{aligned} \nabla_r f &\approx \nabla_r f_0 \\ \nabla_p f &\approx \nabla_p f_0. \end{aligned} \tag{7.17}$$

With this assumption, we can solve eqn. (7.16) and find

$$\delta f = -\tau_m \left\{ \vec{v} \cdot \nabla_r f_0 - q\vec{\mathcal{E}} \cdot \nabla_p f_0 \right\}. \tag{7.18}$$

To evaluate the RHS, let's write the Fermi function as

$$f_0(\vec{p}) = \frac{1}{1 + e^{\Theta}} \tag{7.19a}$$

$$\Theta = \frac{\left(E(\vec{r}, \vec{p}) - F_n(\vec{r})\right)}{k_B T_L} = \frac{\left(E_c(\vec{r}) + E(\vec{p}) - F_n(\vec{r})\right)}{k_B T_L}. \tag{7.19b}$$

(We have replaced the Fermi level, E_F, in the Fermi function with $F_n(\vec{r})$, the quasi-Fermi level or electrochemical potential.) Now we can use the chain rule to write

$$\nabla_r f_0 = \frac{\partial f_0}{\partial \Theta} \nabla_r \Theta \tag{7.20a}$$

$$\nabla_p f_0 = \frac{\partial f_0}{\partial \Theta} \nabla_p \Theta, \tag{7.20b}$$

where

$$\frac{\partial f_0}{\partial \Theta} = k_B T_L \frac{\partial f_0}{\partial E}. \tag{7.21}$$

Using eqns. (7.20) and (7.21) in eqn. (7.18) we find

$$\delta f = \tau_m k_B T_L \left(-\frac{\partial f_0}{\partial E} \right) \left\{ \vec{v} \cdot \nabla_r \Theta - q\vec{\mathcal{E}} \cdot \nabla_p \Theta \right\}. \tag{7.22}$$

Finally, using eqns. (7.20) and (7.19b) to evaluate $\nabla_r \Theta = \nabla_r (E_c - F_n)/k_B T_L$ and $\nabla_p \Theta = \vec{v}/k_B T_L$ recognizing that $-\nabla_r E_c = -q\vec{\mathcal{E}}$, we can write the steady-state, near-equilibrium solution to the BTE as

$$\boxed{\begin{aligned} \delta f &= \tau_m \left(-\frac{\partial f_0}{\partial E} \right) \left\{ \vec{v} \cdot \vec{\mathcal{F}} \right\} \\ \vec{\mathcal{F}} &= -\nabla_r F_n(\vec{r}) + T_L \left[E_c(\vec{r}) + E(\vec{k}) - F_n(\vec{r}) \right] \nabla_r \left(\frac{1}{T_L} \right). \end{aligned}} \tag{7.23}$$

The quantity, $\vec{\mathcal{F}}$, is called the *generalized force* or the *electrothermal field*; it produces deviations from equilibrium. The generalized force consists of two terms, the gradient of the electrochemical potential, F_n, and the gradient of the (inverse) temperature. Recall that in Lecture 4 we saw that $(f_1 - f_2)$ produces current and that differences in Fermi level and temperature lead to a nonzero $(f_1 - f_2)$. In the Landauer approach, there is a clear physical understanding of the generalized force that leads to current flow while in the BTE approach, we find the same result from the mathematical solution.

Equations (7.23) gives the desired solution to the steady-state, near-equilibrium BTE. Now we can evaluate the transport coefficients.

7.4 Transport coefficients

Having solved the BTE, we can now determine various quantities of interest. For example, the 2D, sheet carrier density is

$$n_S(\vec{r}) = \frac{1}{A} \sum_{\vec{k}} \left[f_0(\vec{r}, \vec{k}) + \delta f(\vec{r}, \vec{k}) \right] = \frac{1}{A} \sum_{\vec{k}} f_0(\vec{r}, \vec{k}), \qquad (7.24)$$

where the last expression follows from the fact that δf is odd in momentum, so it integrates to zero. To find the current density in 2D, we evaluate

$$\vec{J}_n(\vec{r}) = \frac{1}{A} \sum_{\vec{k}} (-q)\, \upsilon(\vec{k})\, \delta f(\vec{k}). \qquad (7.25)$$

Note that in this case, only the small component, δf, matters because the product, $\vec{\upsilon} f_0$ is odd in $\vec{k}$ and integrates to zero, while $\vec{\upsilon}\, \delta f$ is even and gives a finite contribution to the integral.

To find the kinetic energy current, we evaluate

$$\vec{J}_W(\vec{r}) = \frac{1}{A} \sum_{\vec{k}} E(\vec{k})\, \upsilon(\vec{k})\, \delta f, \qquad (7.26)$$

and for the heat current,

$$\vec{J}_Q(\vec{r}) = \frac{1}{A} \sum_{\vec{k}} (E - F_n)\, \upsilon(\vec{k})\, \delta f, \qquad (7.27)$$

where $E = E_c(\vec{r}) + E(\vec{k})$ is the total carrier energy. The heat current carried per electron, $(E - F_n)$ is usually established from thermodynamic arguments, but recall from our discussion in Lecture 4 that it is simply the energy that must be absorbed by an electron flowing at the Fermi energy in the contact so that it can enter an energy channel, E, in the device.

If we evaluate the electric and heat currents using the solution, eqn. (7.23), we will get two terms for each of the two currents because the generalized force contains two terms — one for gradients in the electrochemical potential and a second for gradients in the temperature. The resulting four terms are the four thermoelectric transport coefficients. Let's work out one of these four transport coefficients, the sheet conductance, σ_S.

We begin by evaluating the 2D charge current using eqn. (7.25) with the solution, eqn. (7.23) to find

$$
\begin{aligned}
\vec{J}_n(\vec{r}) &= \frac{(-q)}{A} \sum_{\vec{k}} \tau_m \left(-\frac{\partial f_0}{\partial E} \right) \vec{v} \left[\vec{v} \cdot \vec{\mathcal{F}} \right] \\
&= \frac{(-q)}{A} \sum_{\vec{k}} \tau_m \left(-\frac{\partial f_0}{\partial E} \right) (\vec{v}\vec{v}) \cdot \vec{\mathcal{F}} .
\end{aligned}
\tag{7.28}
$$

The quantity, $(\vec{v}\vec{v})$ is a tensor. The BTE makes it relatively easy to treat anisotropic transport for which the transport coefficients become tensors. To keep things simple, however, let's evaluate the x-directed current assuming that F_n varies only in the x direction and that the temperature is uniform. We find

$$
J_{nx}(\vec{r}) = \left\{ \frac{1}{A} \sum_{\vec{k}} q \, v_x^2 \tau_m \left(-\frac{\partial f_0}{\partial E} \right) \right\} \frac{dF_n}{dx} ,
\tag{7.29}
$$

which we can write as

$$
J_{nx} = \sigma_S \frac{d\,(F_n/q)}{dx} ,
\tag{7.30}
$$

where

$$
\sigma_S = \frac{1}{A} \sum_{\vec{k}} q^2 \, v_x^2 \, \tau_m \left(-\frac{\partial f_0}{\partial E} \right) .
\tag{7.31}
$$

To evaluate σ_S, we must evaluate the sum over k-states. This requires a short discussion.

In any finite size material there is a finite number of states. In a small nanostructure, we simply count the states to perform the sum in eqn. (7.31). For larger samples, the states are closely spaced, so we can convert the sum to an integral if we are careful about counting states. The prescription for converting sums to integrals is

$$
\sum_{\vec{k}} (\bullet) \rightarrow \int (\bullet) \, N_k \, d\vec{k} ,
\tag{7.32}
$$

where N_k is the density of states in k-space. We evaluate N_k by applying periodic boundary conditions to the sample, which leads to a set of discrete states uniformly spaced in k-space. See Lundstrom [3, 5] for the derivation. The result is

$$
\begin{aligned}
1D: \quad & N_k \, d\vec{k} = 2 \times \left(\frac{L}{2\pi} \right) dk = \frac{L}{\pi} \, dk \\[1em]
2D: \quad & N_k \, d\vec{k} = 2 \times \left(\frac{A}{4\pi^2} \right) dk_x dk_y = \frac{A}{2\pi^2} \, dk_x dk_y \\[1em]
3D: \quad & N_k \, d\vec{k} = 2 \times \left(\frac{\Omega}{8\pi^3} \right) dk_x dk_y dk_z = \frac{\Omega}{4\pi^3} \, dk_x dk_y dk_z \, .
\end{aligned}
\tag{7.33}
$$

Note that N_k is independent of band structure and that the factor of two accounts for spin degeneracy. With this prescription, we can proceed.

To evaluate eqn. (7.31), we use the 2D prescription, eqns. (7.33),

$$
\begin{aligned}
\sigma_S &= \frac{1}{A} \sum_{\vec{k}} q^2 \, v_x^2 \, \tau_m \left(-\frac{\partial f_0}{\partial E} \right) \\[1em]
&= \frac{1}{A} g_v \frac{A}{2\pi^2} \int_0^\infty \int_0^{2\pi} q^2 v_x^2 \tau_m \left(-\frac{\partial f_0}{\partial E} \right) d\theta k dk \, ,
\end{aligned}
\tag{7.34}
$$

where we have included a valley degeneracy factor, g_v. Using $v_x = v \cos\theta$, the integral over angle can be performed to find

$$
\sigma_S = \frac{g_v q^2}{2\pi} \int_0^\infty v^2 \tau_m(k) \left(-\frac{\partial f_0}{\partial E} \right) k dk \, .
\tag{7.35}
$$

By assuming parabolic energy bands, we find $k dk = (m^*/\hbar^2) dE$ and $v^2 = 2(E - E_c)/m^*$. If we keep things simple by assuming that $\tau_m(E) = \tau_0$ (i.e. a constant scattering time), eqn. (7.35) becomes

$$
\sigma_S = \frac{g_v q^2 \tau_0}{\pi \hbar^2} \int_0^\infty (E - E_c) \left(-\frac{\partial f_0}{\partial E} \right) dE \, ,
\tag{7.36}
$$

which is an integral that we know how to evaluate. The result is

$$
\sigma_S = \frac{g_v q^2 \tau_0 k_B T_L}{\pi \hbar^2} \mathcal{F}_0(\eta_F) \, ,
\tag{7.37}
$$

where $\mathcal{F}_0(\eta_F)$ is the Fermi-Dirac integral of order zero as defined by eqn. (3.20). The final result may not look familiar, but recall that the sheet carrier density is

$$n_S = N_{2D}\mathcal{F}_0(\eta_F) = \left(\frac{g_v m^* k_B T_L}{\pi \hbar^2} \right) \mathcal{F}_0(\eta_F), \qquad (7.38)$$

which can be used to write eqn. (7.37) as

$$\sigma_S = n_s q \left(\frac{q\,\tau_0}{m^*} \right) = n_s\, q\, \mu_n\,. \qquad (7.39)$$

Equation (7.39) is a familiar result, but how does it compare to the Landauer approach? To answer this question, let't go back to eqn. (7.35) and change the variable of integration to energy:

$$\sigma_S = \frac{q^2}{2\pi} \int_0^\infty v^2 \tau_m(E) \left(-\frac{\partial f_0}{\partial E} \right) \left(\frac{g_v m^*}{\hbar^2} \right) dE\,. \qquad (7.40)$$

By recognizing the 2D density-of-states $(D_{2D}(E) = g_v m^* / \pi \hbar^2)$ and re-arranging eqn. (7.40) a bit, we get

$$\sigma_S = \frac{q^2}{2} \int_0^\infty (v\,\tau_m)\, v\, D_{2D}(E) \left(-\frac{\partial f_0}{\partial E} \right) dE\,. \qquad (7.41)$$

Now let's rearrange this expression by inserting factors of $\pi/2$ and $h/4$ and then undoing them with factors of $2/\pi$ and $4/h$. We find

$$\sigma_S = \frac{4}{h} \frac{q^2}{2} \int_0^\infty \left(\frac{\pi}{2} v\,\tau_m \right) \left[\left(\frac{2}{\pi} v \right) \frac{h}{4} D_{2D}(E) \right] \left(-\frac{\partial f_0}{\partial E} \right) dE\,. \qquad (7.42)$$

The first term in parentheses can be recognized as the mean-free-path for backscattering, $\lambda(E)$, (recall eqn. (6.16)). The second term in parentheses is the average velocity at energy, E, in the x direction, $\langle v_x^+ \rangle$. Recall from eqn. (2.25) that the number of conduction channels at energy, E, is $h/4$ times the average velocity in the transport direction times the density-of-states at E, so the term in brackets is $M_{2D}(E)$. We conclude that eqn. (7.42) is

$$\sigma_S = \frac{2q^2}{h} \int_0^\infty \lambda(E) M_{2D}(E) \left(-\frac{\partial f_0}{\partial E} \right) dE\,. \qquad (7.43)$$

The result of solving the Boltzmann Transport Equation in the diffusive limit is exactly the same result obtained from the Landauer approach in the diffusive limit, eqn. (3.57). Similarly, it is easy to show that the BTE gives the same answers as the Landauer approach for the Seebeck coefficient and the electronic heat conductivity. (For a more formal and complete discussion of the relation between the BTE and the Landauer approach, see Jeong *et al.* [6].)

In this section, we learned how to solve the BTE and saw the final result is the same as that obtained from the Landauer approach in the diffusive limit. The advantage of the Landauer approach is its clear connection to the underlying physics. Another advantage is that treating ballistic or quasi-ballistic transport is no more difficult than treating diffusive transport. The BTE, on the other hand, treats anisotropic transport naturally and, as we will discuss next, it is easy to include B-fields in the BTE. It is also useful in treating far from equilibrium transport [3].

Exercise 7.1: Average scattering time, $\langle\langle\tau_m\rangle\rangle$, for power law scattering in 2D

A simple way to describe energy dependent scattering is by assuming a power law form for $\tau_m(E)$,

$$\tau_m = \tau_0 \left(\frac{E - E_c}{k_B T_L}\right)^s, \tag{7.44}$$

where s is a characteristic exponent. Now if we return to eqn. (7.36), the conductivity in 2D, but retain the energy dependence of the scattering time, we have

$$\sigma_S = \frac{g_v q^2}{\pi\hbar^2} \int_0^\infty \tau_m(E)(E - E_c)\left(-\frac{\partial f_0}{\partial E}\right) dE. \tag{7.45}$$

Dividing and multiplying by the sheet carrier density, we find

$$\sigma_S = q^2 \frac{\left(\frac{g_v}{\pi\hbar^2}\right) \int_0^\infty \tau_m(E)(E - E_c)\left(-\frac{\partial f_0}{\partial E}\right) dE}{\int_0^\infty \left(\frac{g_v m^*}{\pi\hbar^2}\right) f_0(E) dE} n_S, \tag{7.46}$$

which can be written as

$$\sigma_S = n_S q \frac{q \langle\langle\tau_m\rangle\rangle}{m^*}, \tag{7.47}$$

where

$$\langle\langle\tau_m\rangle\rangle \equiv \frac{\int_0^\infty \tau_m(E)(E - E_c)\left(-\frac{\partial f_0}{\partial E}\right) dE}{\int_0^\infty f_0(E) dE}. \tag{7.48}$$

Assuming power law scattering, eqn. (7.48) becomes

$$\langle\langle\tau_m\rangle\rangle = \tau_0 \frac{\int_0^\infty \left(\frac{E - E_c}{k_B T_L}\right)^s (E - E_c)\left(-\frac{\partial f_0}{\partial E}\right) dE}{\int_0^\infty f_0(E) dE}. \tag{7.49}$$

The denominator can be integrated to find $k_B T_L \mathcal{F}_0(\eta_F)$. Similarly, the numerator can be integrated to find $k_B T_L \Gamma(s+2)\mathcal{F}_s(\eta_F)$. The result is that eqn. (7.49) becomes

$$\langle\langle \tau_m \rangle\rangle = \tau_0 \frac{\Gamma(s+2)\mathcal{F}_s(\eta_F)}{\mathcal{F}_0(\eta_F)}. \qquad (7.50)$$

(For 1D and 3D transport, similar, but not identical expressions can be derived.) For a nondegenerate semiconductor, the Fermi-Dirac integrals become exponentials, and we find

$$\langle\langle \tau_m \rangle\rangle = \tau_0 \, \Gamma(s+2). \qquad (7.51)$$

Acoustic phonon scattering goes as the density of final states, which is independent of energy in 2D, so $s = 0$. For charged impurity scattering, $s \approx 1$. The result is that in 2D $\langle\langle \tau_m \rangle\rangle$ is one to two times τ_0 for common scattering mechanisms.

7.5   Magnetic fields

When a magnetic field is applied, the carrier transport coefficients change. The use of B-fields in Hall Effect measurements (which will be discussed in Lecture 8) is a common way to characterize the properties of materials. It is straight-forward (though a little tedious mathematically) to solve the BTE with a magnetic field. In this section we will do so for a common experimental condition — the measurement of the conductivity of a 2D sample (in the $x - y$) plane with a B-field applied normal to the sample in the z direction.

We begin with the steady-state BTE in the relaxation time approximation, eqn. (7.16), but assume spatial uniformity to keep things simple, and add a B-field to find

$$-q\vec{\mathcal{E}} \cdot \nabla_p f - q \left(\vec{v} \times \vec{B} \right) \cdot \nabla_p f = -\frac{\delta f}{\tau_m}. \qquad (7.52)$$

We may be tempted to assume (as in eqn. (7.17)),

$$\nabla_p f \approx \nabla_p f_0, \qquad (7.53)$$

but there is a problem. This assumption is fine for the first term on the LHS, but it does not work for the second term. The reason is

$$\nabla_p f_0 = \frac{\partial f_0}{\partial E} \nabla_p E = \frac{\partial f_0}{\partial E} \vec{v}, \qquad (7.54)$$

which, when inserted in the second term on the LHS of eqn. (7.52), gives $\vec{v} \times \vec{B} \cdot \vec{v} = 0$. Instead, to find a solution with a B-field, we must write eqn. (7.52) as

$$-q\vec{\mathcal{E}} \cdot \nabla_p f_0 - q\left(\vec{v} \times \vec{B}\right) \cdot \nabla_p(\delta f) = -\frac{\delta f}{\tau_m}, \qquad (7.55)$$

which is more difficult to solve because δf appears on both sides.

Equation (7.55) can be solved by assuming a form for the solution and solving for the parameter in the assumed form. Let's assume a solution of the form, eqn. (7.23), but with the generalized force, $\vec{\mathcal{F}}$, replaced by an unknown vector, $\vec{G}$, which we must determine. Our assumed solution is

$$\delta f = \tau_m(E) \left(-\frac{\partial f_0}{\partial E}\right) \left\{\vec{v} \cdot \vec{G}\right\}, \qquad (7.56)$$

where $\vec{G}$ is independent of $\vec{p}$. When we insert this assumed solution in eqn. (7.55), we need to evaluate $\nabla_p(\delta f)$. Note from eqn. (7.56), that the assumed solution has a product $\tau_m(E)(-\partial f_0/\partial E)$ out front, but when we take the gradient of a function of energy, the result will be proportional to $\vec{v}$ as in eqn. (7.54), so it will contribute nothing. The result is that we need only to take the gradient of the final term, and we find

$$\nabla_p(\delta f) = \tau_m \left(-\frac{\partial f_0}{\partial E}\right) \frac{\vec{G}}{m^*}, \qquad (7.57)$$

where we have assumed parabolic energy bands ($\nabla_p \vec{v} = 1/m^*$). Now inserting eqns. (7.54), (7.56), and (7.57) in our BTE, eqn. (7.55), we find

$$q\vec{\mathcal{E}} \cdot \vec{v} - \frac{q\,\tau_m}{m^*} \left(\vec{v} \times \vec{B} \cdot \vec{G}\right) + \vec{v} \cdot \vec{G} = 0. \qquad (7.58)$$

Using the properties of a vector triple product, we can write $(\vec{v} \times \vec{B} \cdot \vec{G}) = (\vec{v} \cdot \vec{B} \times \vec{G})$ and then re-express eqn. (7.58) as

$$\vec{v} \cdot \left[-q\vec{\mathcal{E}} + \frac{q\,\tau_m}{m^*} \left(\vec{B} \times \vec{G}\right) - \vec{G}\right] = 0. \qquad (7.59)$$

Since eqn. (7.59) must hold for any $\vec{v}$, the term in brackets must be zero, so

$$\vec{G} = -q\vec{\mathcal{E}} + \frac{q\,\tau_m}{m^*} \left(\vec{B} \times \vec{G}\right). \qquad (7.60)$$

Equation (7.60) is a vector equation for $\vec{G}$; we can solve it by making use of the result,

$$\vec{c} = \vec{a} + \vec{b} \times \vec{c}$$
$$\vec{c} = \frac{\vec{a} + (\vec{b} \times a) + (\vec{a} \cdot \vec{b})\,\vec{b}}{1 + b^2}. \qquad (7.61)$$

Using this result in eqn. (7.60), we find

$$\vec{G} = \frac{-q\vec{\mathcal{E}} - (q^2\tau_m/m^*)\left(\vec{B} \times \vec{\mathcal{E}}\right) - q(q\tau_m/m^*)^2 \left(\vec{\mathcal{E}} \cdot \vec{B}\right)\vec{B}}{1 + (\omega_c\tau_m)^2}, \qquad (7.62)$$

where

$$\omega_c = \frac{qB}{m^*} \qquad (7.63)$$

is the *cyclotron resonance frequency*. In the presence of a magnetic field, electrons orbit at a frequency of ω_c. The plane of the orbit is normal to the direction of the B-field.

Examining eqn. (7.62), we see that the first term in the numerator will give a contribution to the conductivity like eqn. (7.39). We shall see that the second term gives rise to the *Hall effect*. The third term is proportional to B^2 and gives a *magnetoresistance*. The denominator is also quadratic in B and give another contribution to the magnetoresistance.

Now let's consider a simple case, but one that is a common experimental condition — the measurement of the conductivity of a 2D sample (in the $x - y$ plane) with a B-field applied normal to the sample in the z direction (as in Fig. 7.3). Since the electric field lies in the $x - y$ plane, and the B-field is normal to the plane, the last term in eqn. (7.62) is zero. We also assume that the B-field is small, which corresponds to

$$\omega_c\tau_m \ll 1. \qquad (7.64)$$

When the B-field is low, electrons scatter many times before completing an orbit. (Low magnetic fields are a common experimental condition, but interesting things happen under high B-fields, as we will discuss in Lecture 8.) With the assumption of a small B-field, eqn. (7.62) becomes

$$\vec{G} = -q\vec{\mathcal{E}} - (q^2\tau_m/m^*)\left(\vec{B} \times \vec{\mathcal{E}}\right). \qquad (7.65)$$

Since we assume that $\vec{\mathcal{E}}$ lies in the $x - y$ plane and that B is in the z direction, then $\vec{G}$ lies in the $x - y$ plane and has the components

$$\begin{aligned} G_x &= -q\mathcal{E}_x + (q^2\tau_m/m^*)B_z\mathcal{E}_y \\ G_y &= -q\mathcal{E}_y - (q^2\tau_m/m^*)B_z\mathcal{E}_x. \end{aligned} \qquad (7.66)$$

We have solved the BTE in the presence of a magnetic field. The solution is eqn. (7.56) with the vector, $\vec{G}$, given by eqn. (7.62) or by eqn. (7.65) for our assumption of a planar sample with a small, B-field normal to the plane. The analogous solution in the absence of a B-field was eqn. (7.23).

We have assumed that there are no spatial gradients in carrier density or temperature. To include concentration and temperature gradients, simply replace $-q\vec{\mathcal{E}}$ by the generalized force, $\vec{\mathcal{F}}$.

The next step is to see how the transport coefficients change in the presence of a magnetic field, so we need to insert our new solution in eqns. (7.25) and (7.27) and see how the charge and heat currents change. To keep the discussion simple, let's just examine the conductivity.

We begin with eqn. (7.28) with $\vec{\mathcal{F}}$ replaced with $\vec{G}$

$$\vec{J}_n(\vec{r}) = \frac{(-q)}{A} \sum_{\vec{k}} \tau_m \left(-\frac{\partial f_0}{\partial E} \right) \vec{v} \left[\vec{v} \cdot \vec{G} \right] . \tag{7.67}$$

The two components of the current become

$$J_{nx} = \frac{(-q)}{A} \sum_{\vec{k}} \tau_m \left(-\frac{\partial f_0}{\partial E} \right) v_x \left[v_x G_x + v_y G_y \right]$$

$$J_{ny} = \frac{(-q)}{A} \sum_{\vec{k}} \tau_m \left(-\frac{\partial f_0}{\partial E} \right) v_y \left[v_x G_x + v_y G_y \right] . \tag{7.68}$$

Note that for J_{nx}, the second term in the brackets will bring in a sum (integral) over $\vec{k}$ of $v_x v_y = v^2 \cos\theta \sin\theta$, which, when integrated from 0 to 2π will give zero. Similarly, for J_{ny}, the first term in the brackets will bring in a term, $v_y v_x$, which will integrate to zero. Consequently, eqns. (7.68) becomes

$$J_{nx} = \frac{(-q)}{A} \sum_{\vec{k}} \tau_m \left(-\frac{\partial f_0}{\partial E} \right) v_x^2 G_x$$

$$J_{ny} = \frac{(-q)}{A} \sum_{\vec{k}} \tau_m \left(-\frac{\partial f_0}{\partial E} \right) v_y^2 G_y . \tag{7.69}$$

In equilibrium, we have $v_x^2 + v_y^2 = v^2$ and when averaged over angle, $v_x^2 = v_y^2$, so we can replace v_x^2 with $v^2/2$. For parabolic energy bands, $m^* v^2/2 = (E - E_c)$, so the final result is that v_x^2 and v_y^2 in eqns. (7.69) can be replaced with $(E - E_c)/m^*$ to find

$$J_{nx} = \frac{(-q)}{A} \sum_{\vec{k}} \frac{(E - E_c) \tau_m(E)}{m^*} \left(-\frac{\partial f_0}{\partial E} \right) G_x$$

$$J_{ny} = \frac{(-q)}{A} \sum_{\vec{k}} \frac{(E - E_c) \tau_m(E)}{m^*} \left(-\frac{\partial f_0}{\partial E} \right) G_y . \tag{7.70}$$

Finally making use of eqns. (7.66), we find

$$J_{nx} = \frac{1}{A} \sum_{\vec{k}} \frac{(E - E_c)\,\tau_m(E)}{m^*} \left(-\frac{\partial f_0}{\partial E}\right) \{q^2 \mathcal{E}_x - (q^3 \tau_m/m^*) B_z \mathcal{E}_y\}$$

$$J_{ny} = \frac{1}{A} \sum_{\vec{k}} \frac{(E - E_c)\,\tau_m(E)}{m^*} \left(-\frac{\partial f_0}{\partial E}\right) \{q^2 \mathcal{E}_y + (q^3 \tau_m/m^*) B_z \mathcal{E}_x\} \ .$$

$$(7.71)$$

To evaluate the current densities, we divide the RHS of these equations by

$$n_S = \frac{1}{A} \sum_{\vec{k}} f_0(E) \,, \tag{7.72}$$

and multiply by the same factor to obtain

$$\boxed{\begin{aligned} J_{nx} &= \sigma_S \mathcal{E}_x - \sigma_S \mu_H B_z \mathcal{E}_y \\ J_{ny} &= \sigma_S \mu_H B_z \mathcal{E}_x + \sigma_S \mathcal{E}_y \,, \end{aligned}} \tag{7.73}$$

where

$$\sigma_S = n_S \, q \, \mu_n \,, \tag{7.74}$$

and

$$\mu_n = \frac{q \langle\langle \tau_m \rangle\rangle}{m^*} \tag{7.75}$$

with

$$\langle\langle \tau_m \rangle\rangle \equiv \frac{\sum\limits_{\vec{k}} (E - E_c)\,\tau_m(E) \left(-\frac{\partial f_0}{\partial E}\right)}{\sum\limits_{\vec{k}} f_0(E)} \,, \tag{7.76}$$

which is the same as eqn. (7.48). The *Hall mobility* in eqns. (7.73) is defined as

$$\mu_H \equiv \mu_n r_H \,, \tag{7.77}$$

where the *Hall factor is*

$$\boxed{r_H \equiv \frac{\langle\langle \tau_m^2 \rangle\rangle}{\langle\langle \tau_m \rangle\rangle^2}} \,, \tag{7.78}$$

with

$$\langle\langle \tau_m^2 \rangle\rangle \equiv \frac{\sum\limits_{\vec{k}}(E - E_c)\,\tau_m^2(E)\left(-\frac{\partial f_0}{\partial E}\right)}{\sum\limits_{\vec{k}} f_0(E)}. \tag{7.79}$$

Equations (7.73) are desired current equations for a planar sample in the presence of a small B-field normal to the plane. In the absence of the B-field, an x-directed electric field produces only an x-directed current, but in the presence of a normal B-field, it also produces a y-directed current. Similarly, a y-directed electric field produces currents in both x and y directions. In short, the conductivity becomes a tensor. We could follow a similar procedure and work out the corresponding results for the Seebeck coefficient and electronic thermal conductivity, and we would find that they become tensors too [7].

Exercise 7.2: Hall factor for power law scattering in 2D

The Hall factor plays an important role in Hall Effect measurements. For power law scattering, a simple expression for r_H can be easily derived. To evaluate eqn. (7.79), note that

$$\tau_m^2 = \tau_0^2 \left[\left(\frac{E - E_c}{k_B T_L}\right)^s\right]^2 = \tau_0^2 \left(\frac{E - E_c}{k_B T_L}\right)^{2s}, \tag{7.80}$$

so when we evaluate eqn. (7.79), the result will be just like eqn. (7.51) with s replaced by $2s$,

$$\langle\langle \tau_m^2 \rangle\rangle = \tau_0^2 \,\Gamma(2s + 2), \tag{7.81}$$

where we have assumed a non-degenerate semiconductor.

Having evaluated $\langle\langle \tau_m^2 \rangle\rangle$, we can evaluate the Hall factor from eqn. (7.78),

$$r_H \equiv \frac{\langle\langle \tau_m^2 \rangle\rangle}{\langle\langle \tau_m \rangle\rangle^2} = \frac{\Gamma(2s + 2)}{\Gamma(s + 2)^2}. \tag{7.82}$$

As s varies from 0 to 1 (acoustic phonon to charged impurity scattering in 2D), r_H varies from 1 to 1.5. In Hall effect measurements, the precise scattering mechanisms may not be known, so it is common to assume that $r_H = 1$. It is important to remember, however, that this can introduce some uncertainty into the final result.

7.6   Discussion

Equations (7.73) describe an important experimental condition — the measurement of transport in a 2D sample in the presence of a normal magnetic field. You will encounter these equations frequently, since they are widely-used, so we briefly discuss some alternate ways of writing the same equations.

In matrix notation, we can write eqns. (7.73) as

$$\begin{pmatrix} J_{nx} \\ J_{ny} \end{pmatrix} = \begin{bmatrix} \sigma_S & -\sigma_S \mu_H B_z \\ +\sigma_S \mu_H B_z & \sigma_S \end{bmatrix} \begin{pmatrix} \mathcal{E}_x \\ \mathcal{E}_y \end{pmatrix} \qquad (7.83)$$

Alternatively, we can write eqns. (7.73) in vector notation as

$$\vec{J}_n = \sigma_S \vec{\mathcal{E}} - \sigma_S \mu_H \, \vec{\mathcal{E}} \times \vec{B}. \qquad (7.84)$$

The physical significance of these terms is illustrated in Fig. 7.3. If we attempt to force a current in the x direction, it will lead to an electric field in the x direction, and the resulting average velocity of electrons will be in the $-x$ direction. The Lorenz force, $-q\vec{v} \times \vec{B}$, will deflect electrons in the $-y$ direction leading to a y-component of $\vec{J}_n$. If the sample is open-circuited in the y direction, then electrons will pile up on one side, and a deficit of electrons will produce a positive charge on the other side of the sample. The resulting $\mathcal{E}_y$ will produce a force in the $+y$ direction that cancels the Lorenz force in the $-y$ direction. The resulting voltage in the direction transverse to the direction of current flow and B-field is the *Hall voltage*, which we will discuss in Lecture 8.

Equation (7.83) is written in matrix notation. In *indicial notation* it is written as

$$J_{ni} = \sum_j \sigma_{ij} (B_z) \mathcal{E}_j, \qquad (7.85)$$

where the indices, i and j, run over the three coordinate axes, x, y, z or 1, 2, 3. The conductivity matrix, or tensor, σ_{ij}, is a 2×2 matrix, or second rank tensor, defined by eqn. (7.83). Equation (7.85) can also be written in the *summation convention* as

$$J_{ni} = \sigma_{ij} (B_z) \mathcal{E}_j, \qquad (7.86)$$

in which for a repeated index (in this case, j), it is assumed that a summation over the range of the index should be performed.

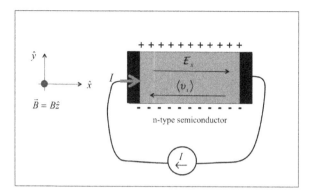

Fig. 7.3. Illustration of a planar sample showing how forcing a current in the x direction in the presence of a z-directed B-field leads to a deflection of electrons in the $-y$ direction.

Equation (7.84) is written in vector notation. We can also write it in indicial notation as

$$J_{ni} = \sigma_S \mathcal{E}_i - \sigma_S \mu_H \epsilon_{ijk} B_k \mathcal{E}_j \,, \qquad (7.87)$$

where we have introduced the *alternating unit tensor*, which is $+1$ when the indices are in cyclic order (e.g. x, y, z), -1 when they are in anti-cyclic order (e.g. y, x, z) and zero otherwise (e.g. x, x, z or y, y, z). In Lecture 8, we will discuss the use of these equations in Hall effect measurements.

7.7 Summary

This lecture has been an introduction to the Boltzmann Transport Equation, which is widely-used to describe low-field transport. For diffusive transport in the absence of a magnetic field, the BTE and Landauer approaches give the same answers. The advantage of the Landauer approach is its physical transparency and the fact that ballistic and quasi-ballistic transport are as easy to handle as diffusive transport. Also note that the basic Landauer transport model does not assume a periodic crystal, so it can, in principle, be applied to amorphous and polycrystalline materials too. The ease with which anisotropic transport and B-fields can be treated are advantages of the BTE approach. The BTE is also useful for treating far from equilibrium transport, a topic not discussed in these lectures. One should be familiar with both approaches, so that the approach best suited to the problem at hand can be used.

7.8 References

The Boltzmann Transport Equation and its solution are discussed in:

[1] N.W. Ashcroft and N.D. Mermin, *Solid–State Physics*, Saunders College, Philadelphia, PA, 1976.

[2] J.M. Ziman, *Principles of the Theory of Solids*, Cambridge Univ. Press, Cambridge, U.K., 1964.

[3] Mark Lundstrom, *Fundamentals of Carrier Transport 2nd Ed.*, Cambridge Univ. Press, Cambridge, UK, 2000.

For an example of how the RTA can fail, see:

[4] D. I. Pikulin, C.-Y. Hou, and C. W. J. Beenakker, Nernst effect beyond the relaxation-time approximation, http://arxiv.org/abs/1105.1303v1, 2011.

Evaluating sums in k-space is discussed in Chapter 1 of [3] and also:

[5] Mark Lundstrom, "ECE-656, Fall 2009, Lecture 2", http://nanohub.org/resources/7281, 2009.

The mathematical relation between the BTE and Landauer approach is discussed in more detail by Jeong, et al.

[6] Changwook Jeong, Raseong Kim, Mathieu Luisier, Supriyo Datta, and Mark Lundstrom, "On Landauer vs. Boltzmann and Full Band vs. Effective Mass Evaluation of Thermoelectric Transport Coefficients", *J. Appl. Phys.*, **107**, 023707, 2010.

For a discussion of how B-fields in the presence of a thermal gradient affect transport leading to the Ettinghausen, Nernst, and Righi-Leduc effects, see:

[7] Charles M. Wolfe, Nick Holonyak, Jr., and Gregory E. Stillman, *Physical Properties of Semiconductors*, Prentice Hall, Englewood Cliffs, New Jersey, 1989.

Lecture 8

Near-equilibrium Transport: Measurements

Contents

8.1 Introduction

Lectures 1-7 have been about the theory of near-equilibrium transport. This lecture is about measuring near-equilibrium transport. These kinds of measurements are widely-used to characterize electronic materials and devices, and the theory we have developed in previous lectures is employed to relate those measurements to the underlying material properties and transport physics. To get accurate results that measure the quantity of interest, electrical characterization must be done carefully. Results can be clouded by several effects, such as contact resistance and uncontrolled thermoelectric effects. Measurements in the absence of a magnetic field are often combined with measurements in the presence of a magnetic field to gain additional information. This lecture is an introduction to some basic measurement considerations.

Since we have four parameters in our coupled current equations, the resistivity (or conductivity), the Seebeck (or Soret) coefficient, the Peltier coefficient, and the electronic heat conductivity, we should measure all four.

Because of the Kelvin relation, however, there is no need to measure the Seebeck and Peltier coefficients separately. The measurement of the Seebeck coefficient (thermopower) is a specialized topic that we will not discuss. Those interested can refer to [1]. The electronic heat conductivity is also difficult to measure, so its value is usually estimated from the Wiedemann-Franz Law. In this lecture, we concentrate on measuring the resistivity (conductivity).

We shall primarily be concerned with diffusive transport for which we can write the current equation as

$$J_{nx} = \sigma_n \frac{d(F_n/q)}{dx} \, . \tag{8.1}$$

The subscript, n, indicates that we are thinking of an n-type material, but similar considerations apply to p-type materials. The differences, when they occur, will be noted. If we assume that the carrier density is uniform, then $d(F_n/q)/dx = \mathcal{E}_x$, and

$$J_{nx} = \sigma_n \mathcal{E}_x \tag{8.2a}$$

$$\mathcal{E}_x = \rho_n J_{nx} \, . \tag{8.2b}$$

We generally measure *conductivity* or *resistivity* because for diffusive samples, these parameters depend on the material, not on the dimensions of the resistor. Equations (8.1) and (8.2) also apply to p-type materials; we simply replace the subscript, n, with p.

Equations (8.1) and (8.2) apply in 1D, 2D, or 3D (we have assumed for simplicity that the spatial variation is only in the x-direction). The units of the current density and resistivity (conductivity) depend on dimensionality, as shown below.

$$
\begin{array}{llllll}
1D: & \mathcal{E}_x & \text{V/m} & J_{nx} & \text{A} & \rho_n & \Omega/\text{m} \\
2D: & \mathcal{E}_x & \text{V/m} & J_{nx} & \text{A/m} & \rho_n & \Omega \\
3D: & \mathcal{E}_x & \text{V/m} & J_{nx} & \text{A/m}^2 & \rho_n & \Omega - \text{m}
\end{array}
\tag{8.3}
$$

In 2D, the resistivity is often called the sheet resistance, ρ_S, and is frequently quoted in units of $\Omega/\square$ — "Ohms per square" because it is the resistance of a square resistor.

Consider the planar resistor shown in Fig. 8.1. The current that flows if a voltage is applied across the two ends is

$$I = GV = \left(\sigma_n \frac{A}{L} \right) V = \left(\sigma_n \frac{Wt}{L} \right) V \, , \tag{8.4}$$

which, using $\sigma_S = (nt)q\mu_n = n_s q\mu_n$, becomes

$$G = \sigma_S \left(\frac{W}{L}\right), \tag{8.5}$$

where $\sigma_S = 1/\rho_S$ is the sheet conductance, and n_S is the sheet carrier density per cm^2. In 3D, $G = \sigma_n A/L$ and in 1D, $G = \sigma_n/L$.

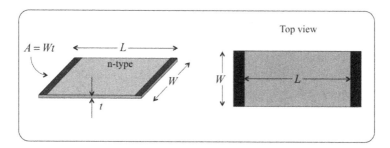

Fig. 8.1. Sketch of a planar resistor with length, L, width, W, and cross-sectional area, $A = Wt$, where t is the thickness of the resistor. If t is large compared to the de Broglie wavelength of electrons, then the electrons in the resistor are three dimensional, but if t is small compared to the de Broglie wavelength, then electrons are quantum mechanically confined in one dimension.

How we compute the sheet conductance depends on whether the electrons are 3D or 2D particles, and that depends on the thickness, t. If t is large enough so that quantum confinement is weak, then we treat the electrons as 3D particles and find the sheet conductance from

$$\sigma_S = \frac{2q^2}{h} \int t\, M_{3D}(E)\lambda(E) \left(-\frac{\partial f_0}{\partial E}\right). \tag{8.6}$$

This is the assumption that we made in getting from eqn. (8.4) to eqn. (8.5), i.e. that $n_S = nt$, where n is the 3D concentration per cm^3. If, on the other hand, t is small so that quantum confinement is strong, then we must treat electrons as 2D particles and σ_S becomes

$$\sigma_S = \frac{2q^2}{h} \int M_{2D}(E)\lambda(E) \left(-\frac{\partial f_0}{\partial E}\right). \tag{8.7}$$

Recall that $M_{3D} = M/A$ is the number of channels per cross sectional area in 3D, and $M_{2D} = M/W$ is the number per width transverse to the direction of current flow in 2D. If the sample is quasi-ballistic, then the mean-free-path, $\lambda(E)$, should be replaced by the apparent mean free path as defined in eqn. (3.40).

Note from eqns. (8.6) or (8.7) that the conductivity depends on the location of the Fermi level, which depends on the carrier density. The conductivity, therefore, is a function of the carrier density, $\sigma_S(n_S)$, so one often measures conductivity vs. carrier density. Having done so, we can then extract the mobility from $\sigma_S \equiv n_S q \,\mu_n$. Finally, we should note that eqns. (8.6) and (8.7) apply to either n- or p-type materials.

8.2   Resistivity/conductivity measurements

Consider the "simple" problem of measuring the resistivity of a material. A planar resistor made with a material of resistivity, ρ_S, is shown in Fig. 8.2. The resistance of the material (i.e. the "channel" of the resistor) is $R_{CH} = \rho_S L/W$, but how do we measure R_{CH}? If we inject a current in contact 2 and extract it from contact 1 and measure the voltage, V_{21}, we find

$$R = \frac{V_{21}}{I} = R_{CH} + 2R_C, \tag{8.8}$$

where R_C is the resistance of the metal-semiconductor contact. The measured resistance includes the contact resistance, so we must either find a way to measure the contact resistance or to do the measurement in a way that removes the effect of the contact resistance. Consider the first approach first.

Figure 8.3 is a sketch of a *transmission line structure*, which consists of a planar resistor with contacts spaced at different distances [2]. If we plot the resistance, $R_{ji} = V_{ji}/I$ versus spacing of the two adjacent contacts, S_{ji},

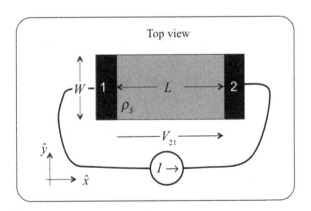

Fig. 8.2.   Sketch of a planar resistor with two metal contacts.

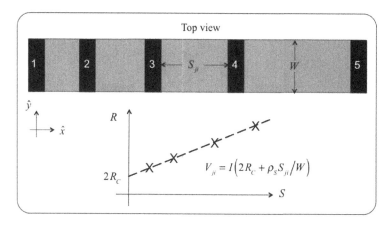

Fig. 8.3.   Sketch of a transmission line structure with a series of differently spaced metal contacts. Also shown is a plot of the resistance between adjacent contacts vs. the spacing between contacts.

we obtain a straight line,

$$R_{ji} = \frac{V_{ji}}{I} = 2R_C + \rho_S \frac{S_{ji}}{W}, \qquad (8.9)$$

as shown in Fig. 8.3. From the slope of the line, we determine ρ_S and from the y-intercept, R_C. Quite a lot can also be learned about the contacts, such as the specific contact resistivity in Ω-m^2 and the so-called "transfer length," the distance which the current penetrates under the metal [2]. Transmission line measurements allow us to characterize both the semiconductor material and the contacts.

Another way to eliminate the contact resistance is to use a *four-probe measurement*, as shown in Fig. 8.4. In this structure, contacts 0 and 5 are called the *current probes*, and contacts 1 and 2 (or 3 and 4) are called the *voltage probes*. A current, I, is injected between contacts 0 and 5 and a voltage, V_{21} is measured between the voltage probes. If a high impedance voltmeter is used, then no current flows in the voltage probes, so there is no voltage drop across the contact resistances. We conclude that

$$R = \frac{V_{21}}{I} = \rho_S \frac{L}{W}, \qquad (8.10)$$

so we measure only the resistance of the material directly with no contribution from the contacts.

Using either of the two approaches, the transmission line structure or the four probe geometry, the resistivity of the material can be measured.

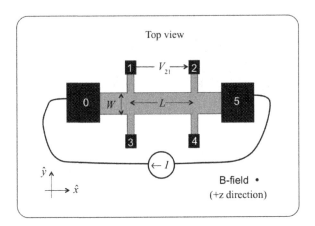

Fig. 8.4.   Sketch of a geomotry used to perform four-probe measurements of resistivity. This is a top view of a structure made of a thin film of material on a substrate. The structure is called a *Hall bar* geometry because, as discussed in the next section, it is also used to perform Hall effect measurements.

For more information about these and related techniques, see [3, 4]. The next step is to measure the carrier density.

8.3   Hall effect measurements

Consider again the Hall bar geometry of Fig. 8.4. If we inject a current in the $+x$ direction, apply a B-field in the $+z$ direction, and measure the resulting voltage in the $+y$ direction (i.e. between contacts 1 and 3 or 2 and 4), then the resulting Hall voltage is positive for an n-type sample and negative for a p-type sample. This is the Hall effect, discovered in 1879 by Edwin Hall and discussed in Chapter 7. It is commonly used to measure the carrier concentration (or, more accurately, a quantity proportional to the carrier concentration) [3, 4].

Figure 8.5 illustrates the essential physics of the Hall effect for an n-type sample. Since current flows in the $+x$ direction, there is an average velocity in the $-x$ direction determined from $I_x = W n_S q \langle v_x \rangle$. The Lorentz force, $\vec{F}_e = -q\vec{v} \times \vec{B}$, produces an average force on electrons in the $-y$ direction. The result is that electrons pile up on the bottom of the sample, and the deficit electrons on the top surface leads to a corresponding positive charge. The resulting electric field in the $-y$ direction balances the force due to the B-field, so that there is no net velocity in the y direction. The corresponding voltage in the y direction, $V_H = -W \mathcal{E}_y$, is the Hall voltage.

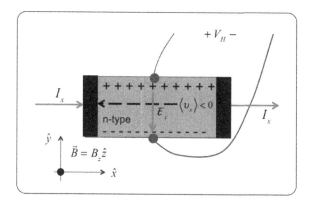

Fig. 8.5. Illustration of the essential physics of the Hall effect.

Hall effect analysis begins with eqn. (7.84)

$$\vec{J}_n = \sigma_n \vec{\mathcal{E}} - (\sigma_n \mu_n r_H)\, \vec{\mathcal{E}} \times \vec{B}\,, \tag{8.11}$$

which can be written as

$$\begin{aligned}
J_{nx} &= \sigma_n \mathcal{E}_x - (\sigma_n \mu_n r_H)\, \mathcal{E}_y B_z \\
J_{ny} &= \sigma_n \mathcal{E}_y + (\sigma_n \mu_n r_H)\, \mathcal{E}_x B_z\,.
\end{aligned} \tag{8.12}$$

Solving these equations for the electric field, we find

$$\begin{aligned}
\mathcal{E}_x &= \rho_n J_{nx} + (\rho_n \mu_n r_H B_z)\, J_{ny} \\
\mathcal{E}_y &= \rho_n J_{ny} - (\rho_n \mu_n r_H B_z)\, J_{nx}\,.
\end{aligned} \tag{8.13}$$

The experimental conditions ensure that $J_{ny} = 0$, so we can solve the second equation for $\mathcal{E}_y$ to find

$$\mathcal{E}_y = -\frac{r_H B_z J_{nx}}{q n_S}\,, \tag{8.14}$$

which can be written as

$$R_H \equiv \frac{\mathcal{E}_y}{J_{nx} B_z} = \frac{r_H}{(-q)n_S} = \frac{1}{(-q)n_H}\,, \tag{8.15}$$

where R_H is known as the *Hall coefficient*, and

$$n_H \equiv \frac{n_S}{r_H} \tag{8.16}$$

is the *Hall concentration*. Recall from eqn. (7.78) that

$$r_H \equiv \frac{\langle\langle \tau_m^2 \rangle\rangle}{\langle\langle \tau_m \rangle\rangle^2}\,, \tag{8.17}$$

is the Hall factor, a number on the order of unity that depends on the precise energy dependence of the scattering mechanisms.

Finally, expressing the Hall coefficient in terms of the experimentally measured voltage, $V_H = -W\mathcal{E}_y$ and injected current, $I = W J_{nx}$, we have

$$\boxed{R_H = \frac{-V_H}{IB_z} = \frac{1}{(-q)n_H}} \,. \tag{8.18}$$

To summarize, in a Hall effect measurement, we apply a z-directed B-field, inject an x-directed current, and measure the resulting voltage in the y direction, the Hall voltage, which is positive for n-type samples and negative for p-type samples. The measured Hall coefficient, R_H, is directly related to the Hall concentration, n_H, according to eqn. (8.16) (for p-type samples, replace $(-q)$ with $(+q)$). The Hall concentration is the "carrier density" usually quoted in Hall effect measurements. It is related to the actual concentration, $n_S = r_H n_H$, by the Hall factor, as given by eqn. (8.17).

Exercise 8.1: Hall effect analysis

To see how things work in practice, let's consider a simple example, a Hall bar measurement with the following conditions:

$$I = 1\,\mu\text{A}$$
$$B_z = 2000 \text{ Gauss or } 0.2 \text{ Tesla}$$
$$L = 100\,\mu\text{m}$$
$$W = 50\,\mu\text{m}$$
$$V_{21} = 0.54 \text{ mV } (B_z = 0)$$
$$V_{24} = 0.13 \text{ mV } (B_z = 0.2 \text{ T})$$

From this information, we can obtain the resistivity, the sheet carrier density (or, rather, the Hall sheet carrier density), and the mobility (actually the Hall mobility).

Consider the resistivity first. The measurement in the absence of a B-field gives

$$R_{xx} = \frac{V_{21}(B_z = 0)}{I} = 540\,\Omega \,. \tag{8.19}$$

Since this resistance is simply related to the sheet resistance by $R_{xx} = \rho_S L/W$, we find

$$\rho_S = 270 \ \Omega/\square \,. \tag{8.20}$$

(As discussed for eqn. (8.3), recall that $\square$ is not a real unit of measurement. The sheet resistance is commonly written as "Ohms per square" because the resistance of a square, $L = W$, resistor is just ρ_S.)

Consider next the concentration. The measured Hall voltage, V_{24}, is positive, so this is an n-type sample. From eqn. (8.18), we find the Hall coefficient to be

$$R_H = \frac{-V_{24}}{IB_z} = -650 \text{ m}^2/\text{C}, \tag{8.21}$$

which, from eqn. (8.18) gives

$$n_H = r_H n_S = 9.6 \times 10^{15} \text{ m}^{-2} = 9.6 \times 10^{11} \text{ cm}^{-2}. \tag{8.22}$$

Finally, having the measured resistivity and Hall carrier density, we can determine the *Hall mobility*. We begin with

$$\sigma_n = \frac{1}{\rho_S} \equiv n_S q \mu_n = n_H q \mu_H, \tag{8.23}$$

where $\mu_H = r_H \mu_n$ as given in eqn. (7.77). Putting numbers in eqn. (8.23), we find

$$\mu_H = 24,100 \text{ cm}^2/\text{V-s}. \tag{8.24}$$

One could, in principle, try to estimate the Hall factor, r_H, and thereby determine the actual carrier density and mobility, but this introduces additional assumptions and uncertainties, so it is common practice to simply quote the Hall concentration and Hall mobility and hope that they are close to the actual concentration and mobility.

8.4 The van der Pauw method

The van der Pauw method provides resistivity and Hall effect measurements on arbitrarily shaped, planar samples without the use of the Hall bar geometry [3–5]. The basic approach is summarized in Fig. 8.6. The sample to be measured lies in the $x - y$ plane. It can be arbitrarily shaped, but it must be homogeneous, conduction must be isotropic, and there can be no holes in the sample. The four contacts are assumed to be small and placed along the perimeter of the sample. To perform a resistivity measurement (Fig. 8.6(a)), we force a current through two adjacent contacts, say M and N, and measure the voltage between the other two, say O and P. The resulting "resistance", $R_{MN,OP} = V_{PO}/I$ is related to the sheet resistance of the sample. To perform a Hall effect measurement (Fig. 8.6(b)), we apply a

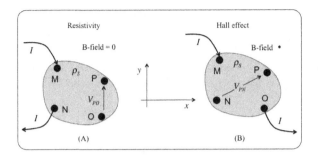

Fig. 8.6.   Illustration of the van der Pauw method. (A) Resistivity measurements and (B) Hall effect measurements.

z-directed B-field, then force a current between two non-adjacent contacts, say M and O, and measure the voltage between the other two, say N and P. The resulting resistance, $R_{MO,NP} = V_{PN}/I$, is related to the Hall voltage.

Since the Hall effect is easier to analyze in this geometry, we begin there and start with eqns. (8.13)

$$\mathcal{E}_x = \rho_S J_{nx} + (\rho_S \mu_n r_H B_z) J_{ny}$$
$$\mathcal{E}_y = \rho_S J_{ny} - (\rho_S \mu_n r_H B_z) J_{nx} \,.$$
(8.25)

The voltage between contact P and N is minus the integral of the electric field along a path that connects the two contacts,

$$V_{PN}(B_z) = - \int_N^P \vec{\mathcal{E}} \cdot d\vec{l} = - \int_N^P (\mathcal{E}_x dx + \mathcal{E}_y dy) \,.$$
(8.26)

If we define the Hall voltage as

$$V_H \equiv \frac{1}{2} \left(V_{PN}(+B_z) - V_{PN}(-B_z) \right),$$
(8.27)

and use eqns. (8.25) in eqn. (8.26), we find

$$V_H = \rho_S \mu_H B_z \left[\int_{y_N}^{y_P} J_{nx} dy - \int_{x_N}^{x_P} J_{ny} dx \right].$$
(8.28)

All of the current injected into contact M must exit contact O, so it must all cross an imaginary line between contacts P and N. Current conservation requires that

$$I = \int_N^P \vec{J}_n \cdot \hat{n} \, dl \,,$$
(8.29)

where $\hat{n}$ is a unit vector normal to the path that connects P and N and is given by

$$\hat{n} \, dl = d\vec{l} \times \hat{z} = dy \, \hat{x} - dx \, \hat{y} \,.$$
(8.30)

Using eqn. (8.30) in eqn. (8.29), we find

$$I = \int_{y_N}^{y_P} J_{nx}\,dy - \int_{x_N}^{x_P} J_{ny}\,dx\,,\tag{8.31}$$

which can be inserted in eqn. (8.28) to find

$$V_H \equiv \frac{1}{2}\left(V_{PN}(+B_z) - V_{PN}(-B_z)\right) = \rho_S\mu_H B_z I\,,\tag{8.32}$$

so Hall effect measurements are readily performed in this geometry.

Consider next how to measure the resistivity in the van der Pauw approach. Fig. 8.7 compares the actual structure (A) with simpler structure (B), an infinite half plane with the four contacts along the lower edge (B). The half plane geometry is much easier to analyze, and van der Pauw showed by a conformal transformation, that the results for geometry (B) are identical to those of geometry (A) providing the material is uniform, isotropic, and contains no holes [5].

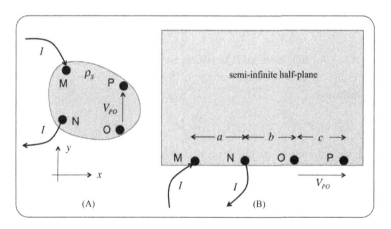

Fig. 8.7. Comparison of the van der Pauw geometry for resistivity measurements (A) with the corresponding measurements on a semi-infinite half plane (B). The results for geometry (B) are identical to those of geometry (A) under some fairly non-restrictive conditions identified by van der Pauw [5].

Consider what happens when we inject a current in contact M of the half-plane. It spreads out radially from the contact, and the 2D current density in A/m is

$$J_{nr} = \frac{I}{\pi r}$$

$$\mathcal{E}_r = \frac{I\rho_S}{\pi r}\,,\tag{8.33}$$

where r is a vector from the contact to the point in question. By integrating eqn. (8.33) from a reference location, r_0, to an arbitrary position, r, we find the potential due to current injected at contact M as

$$V(r) - V(r_0) = -\frac{I\rho_S}{\pi} \ln\left(\frac{r}{r_0}\right).$$ (8.34)

Using this result, we find

$$V(P) = -\frac{I\rho_S}{\pi} \ln\left(\frac{a+b+c}{r_0}\right)$$

$$V(O) = -\frac{I\rho_S}{\pi} \ln\left(\frac{a+b}{r_0}\right)$$ (8.35)

$$V_{PO} = V(P) - V(O) = -\frac{I\rho_S}{\pi} \ln\left(\frac{a+b+c}{a+b}\right)$$

Similarly, there is a contribution with the opposite sign due to the current leaving through contact, N,

$$V'_{PO} = +\frac{I\rho_S}{\pi} \ln\left(\frac{b+c}{b}\right)$$ (8.36)

The total voltage measured between contacts O and P is the sum of the two contributions. When this voltage is divided by the injected current, we get a resistance

$$R_{MN,OP} = \frac{V_{PO} + V'_{PO}}{I} = \frac{\rho_S}{\pi} \ln\left(\frac{(a+b)(b+c)}{b(a+b+c)}\right),$$ (8.37)

which clearly depends on knowing the location of the contacts. Note the order of the subscripts. For $R_{MN,OP}$, the current enters contact M, leaves contact N, and the voltage is measured between contacts O and P with P being the positive terminal. We could also inject the current in contact N, take it out from contact O, and measure the voltage between contact P and M. The resulting resistance would be

$$R_{NO,PM} = \frac{\rho_S}{\pi} \ln\left(\frac{(a+b)(b+c)}{ac}\right),$$ (8.38)

which also depends on the location of the contacts. Finally, eqns. (8.37) and (8.38) can be combined to obtain

$$\exp\left(-\frac{\pi}{\rho_S} R_{MN,OP}\right) + \exp\left(-\frac{\pi}{\rho_S} R_{NO,PM}\right) = 1,$$ (8.39)

which does not depend on the location of the contacts. Although we derived this result for the infinite half plane in Fig. 8.7(b), van der Pauw showed that it also applies to samples like the one in Fig. 8.7(a).

The procedure for measuring resistivity can be summarized as follows. The van der Pauw method can be summarized as follows. First, perform a resistivity measurement (Fig. 8.6(a)) by forcing a current through two adjacent contacts, M and N, and then measuring the voltage between the other two contacts, O and P. The result is $R_{MN,OP} = V_{PO}/I$. Then perform a similar measurement by forcing a current through contacts N and O and measuring the voltage between contacts P and M. The result is a second resistance, $R_{NO,PM} = V_{PO}/I$. Using these two measured resistances, solve eqn. (8.39) for the sheet resistance, ρ_S. Note that the sample does not need to have a special shape, and the contacts do not need to be equally spaced. If the sample is a square, however, then $a = b = c$ and $R_{MN,OP} = R_{NO,PM} = V/I$, so eqn. (8.39) simplifies to:

$$\rho_S = \frac{\pi}{\ln 2} \frac{V}{I}, \qquad (8.40)$$

and there is no need for iteration.

The second part of the van der Pauw measurement is a measurement of the Hall voltage. To perform a Hall effect measurement (Fig. 8.6(b)), apply a z-directed B-field, then force a current between two non-adjacent contacts, M and O, and measure the voltage between the other two, say N and P. Reverse the direction of the B-field and measure the same voltage again. The average of the two measurements is the Hall voltage, as given by eqn. (8.32). From the measured Hall voltage, the Hall coefficient, Hall concentration, and Hall mobility can all be determined.

The Hall bar and van der Pauw geometries provide two different ways to perform the same analysis. It is possible to produce acceptable van der Pauw samples without photolothography and etching, which is its key advantage. If photolithography and etching are available, then either a Hall bar or square van der Pauw sample can be made. The square van der Pauw sample permits a simplified analysis according to eqn. (8.40)

8.5   Temperature-dependent measurements

It is common practice to perform resistivity and Hall effect measurements as a function of temperature, because doing so sheds light on the scattering physics. The results are often presented in the form of mobility vs. temperature plots, as shown in Fig. 8.8(a). (If the mobility was determined from a Hall effect measurement, then what is plotted is actually the Hall mobility.) Typically, the mobility initially increases with temperature, then

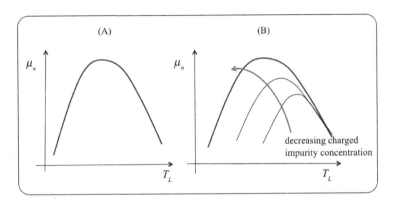

Fig. 8.8.   Sketch of a typical mobility vs. temperature characteristic. (A) General form of the characteristic. (B) Influence of charged impurity scattering on the characteristic.

reaches a plateau and decreases for higher temperatures. The initial increase with temperature is usually attributed to charged (ionized) impurity scattering, and the decrease in mobility with increasing temperature generally indicates phonon scattering. When there are two different scattering mechanisms, the overall mobility is described qualitatively by Mathiessen's Rule (Sec. 4.3.2 in [7]),

$$\frac{1}{\mu_{\text{tot}}} \approx \frac{1}{\mu_{\text{II}}} + \frac{1}{\mu_{\text{ph}}}, \tag{8.41}$$

which states that the lower of the two mobility components controls the overall mobility.

To understand why the mobility increases with temperature, recall Fig. 6.3. Charged impurities introduce fluctuations in the bottom of the conduction band, which act as random scattering potentials. The higher the kinetic energy of carriers (kinetic energy is the distance above the bottom of the conduction band), the less they see these potential fluctuations, so the less they are scattered. The scattering time should increase with energy. Since the average kinetic energy for non-degenerate carriers is proportional to $k_B T_L$, the scattering time (and, therefore the mobility, $\mu_n = q\tau_m/m^*$) should increase with temperature. The precise temperature dependence depends on averaging the energy dependent scattering time, $\tau_m(E - E_c)$ (e.g. as in eqn. (7.48) in 2D). In 3D, ionized impurity scattering leads to a temperature dependence of $T_L^{3/2}$ (see Sec. 4.8 of [7]). The important point is that when the measurements show a mobility that increases with temperature, it generally indicates the presence of charged impurity scattering.

The decrease in mobility at high temperatures is generally due to increased scattering by lattice vibrations (phonons). Thinking of the lattice vibrations as particles, we expect the scattering rate to be proportional to the phonon occupation number,

$$\frac{1}{\tau(E)} \propto n_{\rm ph} . \tag{8.42}$$

Since phonons obey Bose-Einstein statistics, $n_{\rm ph}$ is given by the Bose-Einstein distribution,

$$n_{\rm ph} = \frac{1}{e^{\hbar\omega/k_B T_L} - 1} . \tag{8.43}$$

For covalent semiconductors, acoustic phonon scattering dominates, and one can show by energy-momentum conservation arguments that small energy ($\hbar\omega$) phonons are involved. Accordingly, we find $e^{\hbar\omega/k_B T_L} \approx 1 + \hbar\omega/k_B T_L$, which can be inserted in eqn. (8.43) to find $n_{\rm ph} \approx k_B T_L/\hbar\omega$. (This expression is easy to interpret; it says that dividing the average thermal energy by the energy of a phonon gives the number of phonons.) Finally, from eqn. (8.42) we conclude that $\tau \propto 1/k_B T_L$, so the mobility should decrease with temperature. The precise value of the exponent depends on properly averaging the scattering time over energy. For example, in 3D, we find that $\mu_n \propto T_L^{-3/2}$ when acoustic phonon scattering dominates (Sec. 4.8 of [7]).

For polar semiconductors (e.g. III-V, II-VI semiconductors), the dominant scattering mechanism tends to be polar interactions caused by optical phonons with a frequency, ω_o, where $\hbar\omega_o$ is on the order of $k_B T_L$ at room temperature. In this case, we cannot expand the exponential in the Bose-Einstein distribution for small argument, and must use eqn. (8.43) directly. For polar semiconductors, the mobility decreases exponentially with temperature (Sec. 4.8 of [7]).

Figure 8.8(b) is a sketch of $\mu_n(T_L)$ for three different densities of charged impurities. Beginning at a low temperature, the mobility rises as the temperature increases, and charged impurity scattering becomes less and less effective. The mobility reaches a maximum when the charged impurity scattering equals the phonon scattering. Further increases in temperature cause the mobility to drop as phonon scattering increases. The maximum mobility achieved is limited by charged impurity scattering. The low temperature mobility provides a good measure of the total concentration of charged impurities.

8.6   Discussion

Obtaining reliable data from near-equilibrium transport measurements requires great care; we have simply introduced the techniques in this lecture. More comprehensive texts discuss measurement considerations in detail (e.g. [3, 4]). In this section, we briefly discuss one of the things that can go wrong, because it provides an example for the usefulness of the coupled current equations developed in Lecture 5. We have also restricted our attention to low magnetic fields, but some interesting things happen under high magnetic fields, and we will briefly discuss them in this section too.

A few words about measurement artifacts

Consider again measurements using the Hall bar geometry of Fig. 8.4. The current enters contact 0 and leaves from contact 5. We expect Peltier cooling at contact 5 and Peltier heating at contact 0. We have implicitly assumed isothermal conditions in our analysis of the Hall effect, but what if the heat sink that the Hall bar sits on is not perfect, and there is a temperature gradient in the x direction? How is the measured Hall voltage affected? To answer this question, we should begin with the coupled current equations developed in Lecture 5.

The coupled current eqns. (5.30), can be written in indicial notation as

$$
\begin{aligned}
\mathcal{E}_i &= \rho_{ij}(\vec{B})J_j + S_{ij}(\vec{B})\partial_j T_L \\
J_i^Q &= \pi_{ij}(\vec{B})J_j - \kappa_{ij}^e(\vec{B})\partial_j T_L \,.
\end{aligned}
\tag{8.44}
$$

For an isotropic semiconductor, the four transport tensors are diagonal, but we saw in Lecture 7, that in the presence of a B-field, they acquire off-diagonal components. For low magnetic fields, the extra terms come from the cross product in the Lorentz force, and the diagonal terms are unaffected. For cubic materials, the transport parameters in the presence of a small B-field can be written as

$$
\begin{aligned}
\rho_{ij}(\vec{B}) &= \rho_0 + \rho_0 \mu_H \epsilon_{ijk} B_k + \dots \\
S_{ij}(\vec{B}) &= S_0 + S_1 \epsilon_{ijk} B_k + \dots \\
\pi_{ij}(\vec{B}) &= \pi_0 + \pi_1 \epsilon_{ijk} B_k + \dots \\
\kappa_{ij}^e(\vec{B}) &= \kappa_0^e + \kappa_1^e \epsilon_{ijk} B_k + \dots \,.
\end{aligned}
\tag{8.45}
$$

Each of the four transport coefficients has the same form, a diagonal component denoted with a subscript, 0, and off-diagonal components that involve a cross product described by the alternating unit tensor, ϵ_{ijk} (Sec. 7.6). Except for the resistivity (for which $\rho_1 = \rho_0\mu_H$), we have not specified the off-diagonal component and simply gave them a subscript 1.

Now lets return to the question of how a temperature gradient in the x direction would affect the measured Hall voltage. From eqn. (8.44) we have

$$\mathcal{E}_y = \rho_0 J_y + \rho_0\mu_H\epsilon_{yjk}B_k J_j + S_0\partial_y T_L + S_1\epsilon_{yjk}B_k\partial_j T_L . \qquad (8.46)$$

The experimental conditions dictate that the current is only in the x direction, and the B-field is only in the z direction. Let's also assume that the experimental conditions maintain T_L constant in the y direction. For these conditions, eqn. (8.46) becomes

$$\mathcal{E}_y = +\rho_0\mu_H\epsilon_{yxz}B_z J_x + S_1\epsilon_{yxz}B_z\partial_x T_L . \qquad (8.47)$$

Finally, using the property of the alternating unit tensor, $\epsilon_{yxz} = -1$, we obtain

$$\mathcal{E}_y = -\rho_0\mu_H B_z J_x - S_1 B_z\partial_x T_L . \qquad (8.48)$$

The first term is the Hall effect, which we are trying to measure, but the second term is an artifact that comes from that fact that a temperature gradient in the x direction (produced by the x-directed current being used for the Hall effect measurement) gives rise to an additional component of the y-directed electric field. The y-directed electric field that arises from an x-directed temperature gradient in the presence of a z-directed B-field is called the *Nernst effect*. Note from eqn. (8.48) that if the current and magnetic field were reversed, the first term, the Hall effect term, would not change sign, but the second term would change sign. So by measuring the Hall voltage, then reversing the direction of the current and magnetic field and measuring it again, and averaging the two results, the influence of the Nernst effect could be eliminated (provided that the switch was done quickly so that $\partial T_L/\partial x$ does not change sign). A number of these kinds of *thermomagnetic effects* can occur and may cloud the interpretation of Hall effect measurements [6, 7]. Even in the absence of a B-field, a longitudinal temperature gradient produced by Peltier heating/cooling between contacts 0 and 5 could affect the measured resistivity.

What happens for high magnetic fields?

When solving the BTE in the presence of a B-field, we assumed that the B-field was small, so that

$$\omega_c \tau_m \ll 1\,, \tag{8.49}$$

where for a parabolic energy band, the cyclotron frequency is

$$\omega_c = \frac{qB}{m^*}\,. \tag{8.50}$$

From eqns. (8.49) and (8.50), we can also write the condition for a low magnetic field as

$$\mu_n B \ll 1\,. \tag{8.51}$$

We consider three questions: 1) What does eqn. (8.49) mean physically? and 2) What is the cyclotron frequency for a more general $E(k)$? 3) How do high magnetic fields change the measured results?

Figure 8.9 shows an electron in the $x - y$ plane orbiting a z-directed B-field. The period of the orbit is $T = 2\pi/\omega_c$. The condition, eqn. (8.49), is, therefore, equivalent to $T \gg \tau_m$. For small magnetic fields, the period is much longer than the time between scattering events, so electrons rarely complete an orbit. For high B-field, $T \ll \tau_m$ and electrons typically complete an orbit or orbits before scattering.

The equation of motion for the electron in Fig. 8.9 is

$$\frac{d\left(\hbar \vec{k}\right)}{dt} = -q\vec{v} \times \vec{B}\,. \tag{8.52}$$

Expanding this equation, we find

$$\begin{aligned}
\hbar \frac{dk_x}{dt} &= -qv_y B_z \rightarrow \hbar k \frac{d(\cos\theta)}{dt} = -q\,(v\sin\theta)\,B_z \\
\hbar \frac{dk_y}{dt} &= +qv_x B_z \rightarrow \hbar k \frac{d(\sin\theta)}{dt} = +q\,(v\cos\theta)\,B_z\,.
\end{aligned} \tag{8.53}$$

If the first of these equations is differentiated with respect to time, and then the second equation is used, we find

$$\frac{d^2(\cos\theta)}{dt^2} = -\omega_c^2 \cos\theta\,, \tag{8.54}$$

where

$$\omega_c = \frac{qvB_z}{\hbar k}\,. \tag{8.55}$$

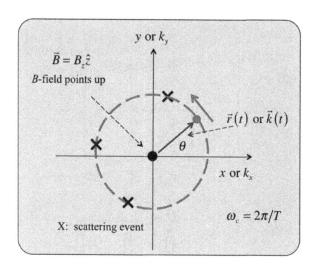

Fig. 8.9. Sketch of an electron in the $x - y$ plane orbiting a z-directed magnetic field. Both the position vector, $\vec{r}(t)$ and the wavevector, $\vec{k}(t)$, undergo a circular orbit.

Equation (8.55) can be used to determine the cyclotron frequency for a general, isotropic energy band. For parabolic bands, $v = \hbar k / m^*$, so $\omega_c = q B_z / m^*$, as expected. In Lecture 10 we will discuss graphene, which has a highly nonparabolic energy band. Equation (8.55) can be used to find the cyclotron frequency of electrons in graphene.

Consider some typical numbers. For lightly doped silicon, the mobility is about 1000 cm^2/V-s. Assume a typical laboratory magnet with a B-field of 0.2 Tesla (2000 Gauss). We find $\mu_n B_z \approx 0.02 \ll 1$, so we are safely in the low magnetic field regime. Magnetic fields that are 10 times higher are not hard to come by, but we would still be in the low magnetic field regime. Magnetic fields that are 100 times higher are needed to operate in the high field regime. Such magnets are available, but they represent the state-of-the-art in magnet technology and are not readily available for routine experiments.

The situation is different for high mobility materials such as III-V semiconductors and modulation-doped heterostructures. For example, consider InGaAs at 300 K. Mobilities on the order of 10^4 are possible. We find $\mu_n B_z \approx 0.2 < 1$. We are still in the low magnetic field regime. If we lower the measurement temperature to 77 K or even lower, however, mobilities of 100,000 or more can be obtained. For these conditions, $\mu_n B_z \approx 2 > 1$. It is, therefore, possible to operate in the high magnetic field regime with

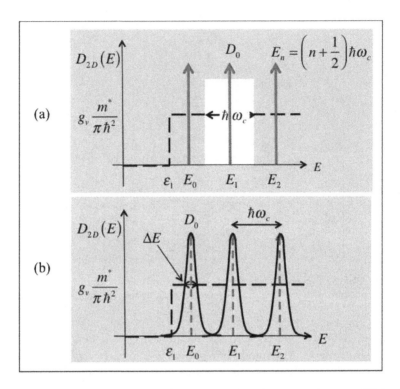

Fig. 8.10. Illustration of how a B-field changes the 2D density-of-states. (A) in the absence of scattering and (B) in the presence of scattering.

commonly available magnets when high mobility samples are used. Interesting things happen at high magnetic fields.

Consider again, eqn (8.54); it describes a harmonic oscillation. We know from quantum mechanics that the energy levels of a harmonic oscillator are quantized according to

$$E_n = \left(n + \frac{1}{2}\right)\hbar\omega_c .\qquad(8.56)$$

In this case, the quantized levels are called *Landau levels*. Consider 2D electrons in the presence of a z-directed B-field. Fig. 8.10(a) illustates how the Landau levels change the density of states. In the absence of the B-field, the density-of-states is

$$D_{2D}(E) = g_v \frac{m^*}{\pi\hbar^2} \qquad (E > \epsilon_1),\qquad(8.57)$$

where ϵ_1 is the bottom of the first subband, and we assume that higher subbands are not occupied. In the presence of a strong B-field, the density-of-states becomes a series of δ-functions,

$$D_{2D}(E, B_z) = D_0 \sum_{n=0}^{\infty} \delta \left[E - \epsilon_1 - \left(n + \frac{1}{2} \right) \hbar \omega_c \right], \qquad (8.58)$$

where D_0 is the degeneracy of each Landau level. Since all of the states must be conserved (they are only re-arranged by the B-field), and since the Landau levels are spaced by $\hbar \omega_c$, we conclude that

$$D_0 = \hbar \omega_c \times \left(g_v \frac{m^*}{\pi \hbar^2} \right) = g_v \frac{2qB_z}{h}. \qquad (8.59)$$

In the presence of scattering the Landau levels broaden. The width of each level is determined from $\Delta E \Delta t \approx \hbar$. Assuming that $\Delta t \approx \tau_m$, the scattering time, then $\Delta E \approx \hbar / \tau_m$. A high magnetic field is defined as a magnetic field high enough so that individual Landau levels can be distinguished. This requires that the spacing of Landau levels, $\hbar \omega_c$, be greater than the spread, ΔE. This leads to the condition that $\omega_c \tau_m \gg 1$ as the condition for a high magnetic field, and gives us another physical interpretation eqn. (8.49).

Exercise 8.2: Shubnikov-de Haas oscillations

If Hall effect and resistivity measurements are done as a strong magnetic field is varied, interesting things can happen. First, consider some numbers. Assume a modulation-doped semiconductor film with $n_S = 5 \times 10^{11}$ cm^{-2} and a mobility of 1×10^5 cm^2/V-s. If $B_z = 1$ T, how many Landau levels are occupied?

First, we should determine whether separate Landau levels can be distinguished. The mobility of 1×10^5 cm^2/V-s $= 10$ m^2/V-s, so

$$\mu_n B_z = 10 \gg 1, \qquad (8.60)$$

and we are clearly in the high B-field regime. Next, we should compute the degeneracy of the Landau levels. Assuming $g_v = 1$, we find

$$D_0 = \frac{2qB_z}{h} = 4.8 \times 10^{10} \text{cm}^{-2},$$

so we find that

$$\frac{n_s}{D_0} = 10.4.$$

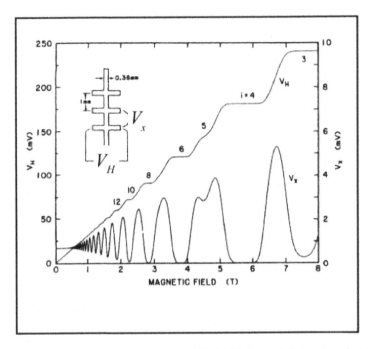

Fig. 8.11.   Hall effect measurements for an AlGaAs/GaAs modulation doped structure at 1.2 K in the presence of a large magnetic field. The injected current is 25.5 mA and the 2D sheet carrier density is 5.6×10^{11} cm^{-2}. (From [8]. Copyright 1985 IEEE. Reprinted, with permission, from *IEEE Transactions on Instrumentation and Measurement.*)

There are 10.4 Landau levels occupied; the first ten are completely filled, and the Fermi level lies within the eleventh Landau level.

Figure 8.11 shows a typical measurement for such a structure [8]. Consider first the longitudinal resistance ($R_{xx} = V_x/I$). For low fields, it is independent of the B-field, and is simply related to the resistivity of the sample, as we have discussed. As the B-field increases, R_{xx} begins to oscillate. This occurs as the Landau levels become well defined. When the Fermi level lies within a Landau level, R_{xx} is low, but as the B-field increases, the spacing between Landau levels increases, the degeneracy of each Landau level increases, and the position of the Fermi level changes. When the Fermi level lies between Landau levels, there are few states, and R_{xx} increases. The oscillations in R_{xx} are known as *Shubnikov-de Haas (SdH) oscillations* and from the period of the oscillation, the carrier density can be measured [9]. When the B-field is very high, R_{xx} actually goes to zero! This is the *quantum Hall Effect* [8 -11].

The Hall voltage in Fig. 8.11 is also interesting. For small B-fields, the Hall voltage is proportional to B, as expected from eqn. (8.32). As the B field increases, however, we see that the Hall voltage becomes quantized and begins to increase in steps. The steps in V_H align with the zero resistance regions in R_{xx}. The quantization in the Hall voltage can be so precise that it can be used as a resistance standard [8].

A satisfactory discussion of the quantum Hall effect would take us too far afield for this lecture. Interested readers should consult [9-11].

8.7 Summary

This lecture has been a short introduction to some techniques commonly used to characterize near-equilibrium transport. Hall bar or van der Pauw geometries allow measurement of both the resistivity and the Hall concentration, from which the Hall mobility can be deduced. In some cases, one can estimate the value of the Hall factor, r_H, in order to estimate the true carrier density and mobility, but it is common practice to simply quote the Hall concentration and Hall mobility. Temperature dependent measurements help experimentalists identify the dominant scattering mechanisms. Care must be taken to exclude galvano and thermomagnetic effects like the Nernst effect, the Righi-Leduc effect, and the Ettingshausen effect, which can affect Hall effect measurements. Finally, high magnetic fields provide additional information, but to operate in the high magnetic field regime with commonly available magnets, high mobility samples are required.

8.8 References

Measurement of the Seebeck coefficient (thermopower) is not discussed in this lecture. For a starting point (as well as a discussion of measuring resistivity), see:

[1] Lee Danielson, "Measurement of the Thermeoelctric Properties of Bulk and Thin Film Materials", Short Courese on Thermoelectrics, SCTs-96, International Conference on Thermoelectrics, Pasadena, CA, March 25, 1996. http://www.osti.gov/bridge/servlets/purl/663573-S3SuWo/webviewable/663573.pdf

The transmission line method is widely used to measure sheet resistance and to characterize contacts. The original reference on the method is:

[2] H.H. Berger, "Models for Contacts to Planar Devices", *Solid-State Electron.*, **15**, 145-158, 1972.

For a comprehensive survey of techniques for characterizing electronic materials and devices (including, but not limited to near-equilibrium electronic transport), see

[3] D.K. Schroder, *Semiconductor Material and Device Characterization*, 3rd Ed., IEEE Press, Wiley Interscience, New York, 2006.

Look focuses on electrical characterization of GaAs, but the techniques he discusses are broadly applicable.

[4] D.C. Look, *Electrical Characterization of GaAs Materials and Devices*, John Wiley and Sons, New York, 1989.

The original paper on the widely-used van der Pauw technique is:

[5] L.J. van der Pauw, "A method of measuring specific resistivity and Hall effect of discs of arbitrary shape", *Phillips Research Reports*, **13**, pp. 1-9, 1958.

For a discussion of how B-fields in the presence of a thermal gradient affect transport leading to the Ettinghausen, Nernst, and Righi-Leduc effects, see:

[6] Charles M. Wolfe, Nick Holonyak, Jr., and Gregory E. Stillman, *Physical Properties of Semiconductors*, Prentice Hall, Englewood Cliffs, New Jersey, 1989.

[7] Mark Lundstrom, *Fundamentals of Carrier Transport 2nd Ed.*, Cambridge Univ. Press, Cambridge, UK, 2000.

The use of quantum Hall effect measurements as a resistance standard is discussed by Cage, et al.

[8] M.E. Cage, R.F. Dziuba, and B.F. Field, "A Test of the Quantum Hall

Effect as a Resistance Standard", *IEEE Trans. Instrumentation and Measurement*, **IM-34**, pp. 301-303, 1985.

To learn about the Shubnikov-de Hass and quantum Hall effects, consult:

[9] D.F. Holcomb, "Quantum electrical transport in samples of limited dimensions", *Am. J. Phys.*, **67**, pp. 278-297, 1999.

[10] John H. Davies, *The Physics of Low-Dimensional Semiconductors*, Cambridge Univ. Press, Cambridge, UK, 1998.

[11] Supriyo Datta, *Electronic Transport in Mesoscopic Systems*, Cambridge Univ. Press, Cambridge, UK, 1995.

Phonon Transport

Contents

9.1 Introduction

We have seen that electrons produce both charge and heat currents. In metals electrons carry most of the heat, but in semiconductors, electrons carry only a part of the heat; most of the heat is carried by lattice vibrations, or *phonons*.

The heat flux due to phonons is

$$J_Q = -\kappa_L dT_L/dx \quad \text{W/m}^2 . \tag{9.1}$$

Our goal in this lecture is to understand what controls the *lattice thermal conductivity*, κ_L, and how it is related to the basic materials parameters. We will see that the same techniques used to describe electron transport can also be used for phonon transport. From eqn. (9.1), we see that the units of κ_L are W/m-K. An exceptional thermal conductor, like diamond, has a thermal conductivity of about 2000 W/m-K while a poor thermal conductor like glass has a thermal conductivity on the order of 1 W/m-K and air of 0.025 W/m-K. Electrical conductivities of solids vary over more than 20 orders of magnitude, but thermal conductivities of solids vary over a range of only 3-4 orders of magnitude.

This lecture is a brief introduction to thermal transport by phonons. The goal is to provide a starting point for learning about phonon transport. Another goal is to discuss the similarities and differences between electron and phonon transport.

9.2   Electrons and phonons

Electrons in the conduction band of a semiconductor feel the influence of the crystal potential and the potential of other electrons. This complicated many body system can be described in most cases as if it consisted of weakly interacting particles. These fictitious electron "quasi-particles" travel through the solid much like electrons travel in a vacuum, but they obey a modified version of Newton's Laws (i.e. their effective mass is different from the rest mass of an electron in vacuum).

In quantum mechanics, electrons can be treated as waves or as particles. The *dispersion* of the electron waves is a plot of the allowed values of an electron's frequency, ω, vs. its wavevector, $\vec{k}$. Instead of plotting $\omega(\vec{k})$ for electrons, we typically plot $E(\vec{k}) = \hbar\omega(\vec{k})$. Figure 9.1(a) is a sketch of a typical dispersion for electrons. Because the crystal lattice is periodic in real space, the dispersion is periodic in k-space, and all unique solutions reside within a *Brillouin zone*. Note also that each band has a finite *bandwidth* — the range of energies occupied by the band. To describe an electron quasi-particle, we form a *wavepacket*, a superposition of electron waves with different wavevectors. The velocity of the electron is the *group velocity* of the wavepacket,

$$\vec{v}_g(\vec{k}) = \frac{1}{\hbar}\nabla_k E(\vec{k}),\qquad(9.2)$$

so the slope of $E(\vec{k})$ at $\vec{k} = \vec{k}_0$ tells us the velocity of an electron wavepacket centered at $\vec{k}_0$. Also shown in Fig. 9.1(a) (dashed line) is the effective mass approximation, which is widely-used to describe $E(\vec{k})$ near the bottom (or top) of a band.

The vibrations of a crystal lattice also comprise a complicated, interacting system that can be described in most cases as if it consists of weakly interacting particles. For lattice vibrations, the quasiparticles are *phonons*. The dispersion of the lattice vibrations is a plot of the allowed values of the frequency, ω, vs. wavevector, $\vec{q}$. Figure 9.1(b) is a sketch of a typical

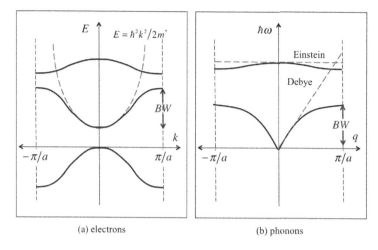

(a) electrons   (b) phonons

Fig. 9.1.   Sketch of dispersions for (a) electrons and (b) phonons. Simplified dispersions commonly used for analytical calculations are also shown as dashed lines. See Fig. 9.3 for examples of realistic dispersions in silicon.

dispersion for phonons. Because the crystal lattice is periodic in real space, the phonon dispersion is periodic in q-space, and all unique solutions reside within the same Brillouin zone that describes the electron dispersion. Note also that each band has a finite bandwidth — just as for electrons.

To describe a phonon as a particle, we form a wavepacket, a superposition of lattice vibrations with different wavevectors. The velocity of the phonon is the *group velocity* of the wavepacket,

$$\vec{v}_g(\vec{q}) = \nabla_q \omega(\vec{q}) , \qquad (9.3)$$

so the slope of $\omega(\vec{q})$ at $\vec{q} = \vec{q_0}$ tells us the velocity of a phonon wavepacket centered at $\vec{q_0}$. Also shown in Fig. 9.1(b) (dashed lines) are some common approximations widely-used to describe $\omega(\vec{q})$. The *Debye approximation* describes the low frequency modes by a linear dispersion ($\omega(q) = v_s q$), and the *Einstein approximation* describes the high frequency modes by $\omega(q) = \omega_0$.

We can appreciate a little about lattice vibrations by thinking of the bonding forces between atoms as springs holding masses together. To first order, the potential is *harmonic*, $U(r) = 1/2 k_s (r - r_0)^2$, where k_s is the spring constant, and r_0 is the rest position. We learn in freshman physics that a mass, M, on a spring will oscillate at a frequency, $\omega = \sqrt{k_s/M}$. Classically, a harmonic oscillator can have any energy, but quantum

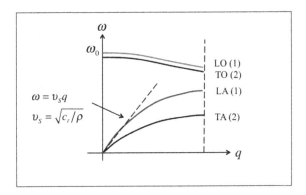

Fig. 9.2. Sketch of a typical phonon dispersion showing the longitudinal acoustic (LA) and optical (LO) modes and the transverse acoustic (TA) and optical (TO) modes.

mechanically, we know that the energy must be quantized according to

$$E_n = \left(n + \frac{1}{2}\right)\hbar\omega\,,\qquad(9.4)$$

where n is an integer, and $E_0 = \hbar\omega/2$ is the *zero point energy*. Instead of talking about the nth excited state of the normal mode with frequency, ω, we will think of n as the number of phonons, each with an energy, $\hbar\omega$.

The sketch in Fig. 9.2 provides a little more detail about phonon dispersion. Recall that for electrons, we have two spin states. For three-dimensional solids, there are three polarization states for lattice vibrations, one for atoms displaced in the direction of propagation (longitudinal) and two for atoms displaced orthogonal to the direction of propagation (transverse). The low energy modes are called *acoustic* modes. There are three acoustic modes, one longitudinal (LA, analogous to sound waves that propagate in air) and two transverse acoustic (TA) modes. Near $q = 0$, acoustic modes display a linear dispersion. For the longitudinal mode,

$$\begin{aligned}\omega(q) &= v_s q\\ v_s &= \sqrt{c_l/\rho}\,,\end{aligned}\qquad(9.5)$$

where v_s is the sound velocity. The elastic constant, c_l, plays the role of the spring constant and ρ of the mass. Typical sound velocities are on the order of 5×10^5 cm/s, about 20 times slower that the velocity of a typical electron ($\approx 10^7$ cm/s). Equation (9.5) shows that materials with heavier atoms have lower sound velocities and correspondingly lower dispersion bandwidths than materials with lighter atoms.

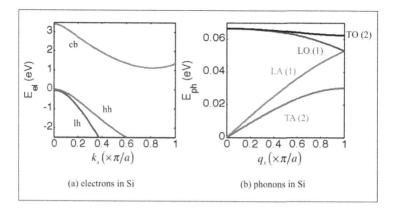

Fig. 9.3.   Realistic computed dispersions along a [100] direction in silicon. (a) electrons and (b) phonons. (Electron dispersion from Band Structure Lab, A. Paul, *et al.*, 2011, DOI: 10254/nanohub-r1308.18. Phonon dispersion after Jeong, *et al.* [5]).

Also shown in Fig. 9.2 are the three optical modes. These modes display little dispersion; ω is almost independent of q. The difference between acoustic and optical modes is that near $q = 0$, adjacent atoms are displaced in the same direction for acoustic modes but in opposite directions for optical modes. In polar materials, the optical modes can interact with light, which is the origin of their name. Because the phonon velocity is given by the slope of $\omega(q)$, the acoustic modes have a relatively high velocity, and the optical modes a low velocity. We expect, therefore, to find that the acoustic modes transport most of the heat.

Figure 9.3 is a comparison of realistic, computed dispersions for silicon along a [100] direction. Shown in Fig. 9.3(a) is the electron dispersion showing the light and heavy hole valence bands and the conduction band. Note that the maximum of the valence band is at $k = 0$, but the minimum of the conduction band is near the zone boundary. Silicon is an indirect semiconductor.

Figure 9.3(b) shows the phonon dispersion for silicon with the three acoustic and three optical branches. In a nonpolar material like Si, the LO and TO branches are degenerate at $q = 0$ and the LO and LA branches are degenerate at the zone boundary. The energy of the optical modes is greater than $k_B T_L$ at 300 K, so the population of the optical modes is relatively small at room temperature and below. The most important thing to note in comparing Figs. 9.3(a) and 9.3(b) is the difference in the bandwidths. The BWs of the conduction and valence bands are $\gg k_B T_L$ at 300 K while the BWs of the phonon dispersion are on the order of $k_B T_L$. In practice, this means that only electron states near the top of the valence band and bottom of the conduction band are important, while for phonons, most of the states in the acoustic branches are occupied. This leads to a significant difference in the wavelength of average electron and phonon.

To estimate the wavelength of an average electron, let's assume a non-degenerate, 3D, semiconductor for which

$$\langle E \rangle = 3k_B T_L / 2 \,. \tag{9.6}$$

From the effective mass approximation, we find for an average electron,

$$\langle E \rangle = \frac{\hbar^2 \left\langle k^2 \right\rangle}{2m^*} \,. \tag{9.7}$$

By equating the above two expressions, we can find an expression for the rms average wavevector, $\sqrt{k^2} = 2\pi / \left\langle \lambda_B^{\text{el}} \right\rangle$, where $\left\langle \lambda_B^{\text{el}} \right\rangle$ is the average electron de Broglie wavelength. The result is

$$\left\langle \lambda_B^{\text{el}} \right\rangle = \frac{h}{\sqrt{3m^* k_B T_L}} \approx 60 \ \overset{\circ}{\text{A}} \,, \tag{9.8}$$

where we assumed $m^* = m_0$ and $T_L = 300$ K.

To estimate the wavelength of an average LA phonon, we recognize that the average thermal energy of the phonons is also given by eqn. (9.6). Instead of eqn. (9.7) for the dispersion, we use the Debye approximation to write

$$\langle E \rangle = \hbar \upsilon_s \langle q \rangle \,. \tag{9.9}$$

By equating eqn. (9.6) to (9.9) and using $\langle q \rangle = 2\pi / \left\langle \lambda_B^{\text{ph}} \right\rangle$, where $\left\langle \lambda_B^{\text{ph}} \right\rangle$ is the average phonon de Broglie wavelength, we find

$$\left\langle \lambda_B^{\text{ph}} \right\rangle = \frac{4\pi \hbar \upsilon_s}{3k_B T_L} \approx 5 \ \text{A} \,, \tag{9.10}$$

where we have again assumed $T_L = 300$ K.

These simple calculations show that the average wavelength of a phonon is about ten times less than that of an average electron. The difference can be traced back to the different bandwidths of the dispersions for electrons and phonons. For electrons, the thermal energy permits only states near the bottom of the conduction band (where the wavevector is small and wavelength long) to be occupied. For phonons, however, the thermal energy is sufficient to occupy most of the acoustic branch, so there is a wide range of wavevectors. The average wavevector is much larger than for electrons, so the average wavelength of a phonon is much shorter.

This section has been a very brief introduction to phonons covering just the essentials needed for this lecture. For a thorough treatment of both electron and phonon dispersions, see [1-4].

9.3 General model for heat conduction

In Lecture 2, we developed a general model for current flow and expressed it in eqn. (2.46) as

$$I = \frac{2q}{h} \int T_{el}(E) M_{el}(E) (f_1 - f_2) \, dE \quad \text{(A)}. \qquad (9.11)$$

The current is proportional to the probability that electrons can transmit through a channel at energy, E, $T_{el}(E)$, times the number of electron channels at that energy, $M_{el}(E)$, times the difference in Fermi functions of the two contacts, $f_1(E) - f_2(E)$.

Figure 9.4 is a sketch of a device that involves phonon transport. We seek an expression for the heat flow through a channel, a material characterized by its dispersion. The channel may be a bulk material, or a nanostructure with reduced dimensionality and a dispersion much different from the bulk. At the two ends of the channel are large thermal reservoirs that maintain thermal equilibrium distributions of phonons at two different temperatures. For electrons, the states in the reservoir were filled according to the equilibrium Fermi function,

$$f_0(E) = \frac{1}{e^{(E-E_F)/k_B T_L} + 1}, \qquad (9.12)$$

but phonons obey Bose statistics, and phonon states are filled according to the equilibrium Bose-Einstein distribution,

$$n_0(\hbar\omega) = \frac{1}{e^{\hbar\omega/k_B T_L} - 1}. \qquad (9.13)$$

For the left contact, $T_L = T_{L1}$ and for the right contact, $T_L = T_{L2}$. Just as for electrons, we assume ideal, absorbing contacts. We assume that phonons that enter a contact are not able to reflect back into the device, so the transmission, T_{ph}, describes the transmission of phonons across the channel.

It is now easy to generalize eqn. (9.11) and apply it to heat transport. For the phonon heat current, we replace the electron energy, E, with the phonon energy, $\hbar\omega$. For the electric current, we have q, the charge carried by an electron. For the heat current, we replace q with $\hbar\omega$, the energy carried

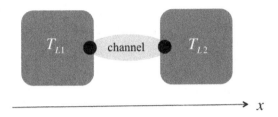

Fig. 9.4.   Sketch of a device for which we seek the heat flow from contact 1 (left) to contact 2 (right).

by a phonon, and move it inside the integral. Finally, the 2 in eqn. (9.11) refers to the two spin polarizations of the electron. For phonons, we remove the 2 and simply absorb the number of polarization states that contribute to heat flow in $M_{\rm ph}$. The final expression for the heat current due to phonons, analogous to the electric current given by eqn. (9.11), is

$$Q = \frac{1}{h} \int (\hbar\omega)\, T_{\rm ph}(\hbar\omega) M_{\rm ph}(\hbar\omega)\, (n_1 - n_2)\, d(\hbar\omega)\ \ ({\rm W}).$$ (9.14)

Since we are interested in near-equilibrium heat transport, we now need to simplify eqn. (9.14) for small differences in temperature. If $T_{L1} \approx T_{L2}$, then $n_1 \approx n_2$, and we can expand n_2 in a Taylor series to find

$$n_2 \approx n_1 + \frac{\partial n_0}{\partial T_L} \Delta T_L\,,$$ (9.15)

which leads to

$$(n_1 - n_2) \approx -\frac{\partial n_0}{\partial T_L} \Delta T_L\,.$$ (9.16)

The derivative, $\partial n_0/\partial T_L$ is readily evaluated from eqn. (9.13). It's also easy to evaluate the derivative, $\partial n_0/\partial(\hbar\omega)$ and to show

$$\frac{\partial n_0}{\partial T_L} = \frac{\hbar\omega}{T_L}\left(-\frac{\partial n_0}{\partial(\hbar\omega)}\right),$$ (9.17)

where

$$\frac{\partial n_0}{\partial(\hbar\omega)} = \left(-\frac{1}{k_B T_L}\right) \frac{e^{\hbar\omega/k_B T_L}}{\left(e^{\hbar\omega/k_B T_L} - 1\right)^2}\,.$$ (9.18)

Finally, putting this all together and inserting it in eqn. (9.14), we find for small differences in temperature,

$$Q = -K_L \Delta T_L\,,$$ (9.19)

where K_L is the thermal conductance in W/K and is given by

$$K_L = \frac{k_B^2 T_L}{h} \int T_{\text{ph}}(\hbar\omega) M_{\text{ph}}(\hbar\omega) \left\{ \left(\frac{\hbar\omega}{k_B T_L} \right)^2 \left(-\frac{\partial n_0}{\partial(\hbar\omega)} \right) \right\} d(\hbar\omega) \,.$$

(9.20)

Equation (9.19) is just Fourier's Law for heat flow; it says that heat flows down a temperature gradient. The thermal conductance displays some similarities to the electrical conductance that we should discuss.

Recall from Lecture 2, eqn. (2.51) that the electrical conductance is

$$G = \frac{2q^2}{h} \int T_{\text{el}}(E) M_{\text{el}}(E) \left(-\frac{\partial f_0}{\partial E} \right) dE \,.$$

(9.21)

The term,

$$W_{\text{el}}(E) = (-\partial f_0 / \partial E) \,,$$

(9.22)

acts as a "window function" that determines which channels carry the electrical current. Note that the electron window function is normalized,

$$\int_{-\infty}^{+\infty} \left(-\frac{\partial f_0}{\partial E} \right) dE = 1 \,.$$

(9.23)

In looking at eqn. (9.20), we see that the term in brackets seems to be acting as a window function to determine which modes carry the heat current, but if we integrate this function,

$$\int_{0}^{+\infty} \left(\frac{\hbar\omega}{k_B T_L} \right)^2 \left(-\frac{\partial n_0}{\partial(\hbar\omega)} \right) d(\hbar\omega) = \frac{\pi^2}{3} \,,$$

(9.24)

we see that it is not normalized. We can define a normalized phonon window function as

$$W_{\text{ph}}(\hbar\omega) \equiv \frac{3}{\pi^2} \left(\frac{\hbar\omega}{k_B T_L} \right) \left(\frac{\partial n_0}{\partial(\hbar\omega)} \right)$$

(9.25)

and write the thermal conductance as

$$K_L = \frac{\pi^2 k_B^2 T_L}{3h} \int T_{\text{ph}}(\hbar\omega) M_{\text{ph}}(\hbar\omega) W_{\text{ph}}(\hbar\omega) d(\hbar\omega) \,,$$

(9.26)

where, as we will see later, $(\pi^2 k_B^2 T_L / 3h)$ is the quantum of thermal conductance. Comparing eqns. (9.21) and (9.26) we see that the electrical and thermal conductances are very similar. Each is proportional to a quantum of conductance times an integral of the transmission times the number of modes times a window function.

The window functions for electrons and phonons are plotted in Fig. 9.5. The two window functions have similar shapes, and each has a width of

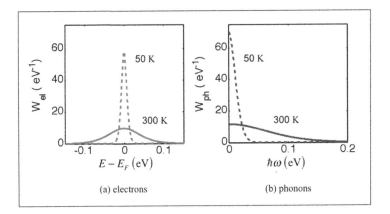

Fig. 9.5.   Plot of the window functions for electrons and phonons. Solid lines: 300 K and dashed lines: 50 K. (a) Electron window function as given by eqn. (9.22) and (b) phonon window function as given by eqn. (9.25). For electrons, the abscissa has both positive and negative values because the energy, $E - E_F$ can be positive or negative. For phonons, the abscissa is only positive because the phonon energy, $\hbar\omega$ is always greater than zero.

a few $k_B T_L$. Along with the distribution of modes as determined by the dispersion, these two window functions play a key role in determining the electrical and thermal conductances.

9.4   Thermal conductivity

The thermal conductivity of a large, diffusive sample is a key material parameter that controls performance in many applications. For a diffusive sample,

$$T_{\text{ph}}(\hbar\omega) = \frac{\lambda_{\text{ph}}(\hbar\omega)}{\lambda_{\text{ph}}(\hbar\omega) + L} \rightarrow \frac{\lambda_{\text{ph}}(\hbar\omega)}{L}. \tag{9.27}$$

We also recognize that for large, 3D samples,

$$M_{\text{ph}} \propto A, \tag{9.28}$$

where A is the cross-sectional area of the structure.

Returning now to eqn. (9.19), we can multiply and divide the right-hand-side by A/L to find

$$Q = -\left(K_L \frac{L}{A}\right) A \frac{\Delta T_L}{L}. \tag{9.29}$$

If we divide both sides by A and recognize that $\Delta T_L/L = dT_L/dx$, we find

$$J_Q = \frac{Q}{A} = -\kappa_L \frac{d\Delta T_L}{dx}, \qquad (9.30)$$

where

$$\kappa_L = K_L \frac{L}{A} \ (\text{W/m-K}). \qquad (9.31)$$

Finally, using eqns. (9.26), (9.27), and (9.31), we find the lattice thermal conductivity as

$$\boxed{\kappa_L = \frac{\pi^2 k_B^2 T_L}{3h} \int \frac{M_{\mathrm{ph}}(\hbar\omega)}{A} \lambda_{\mathrm{ph}}(\hbar\omega) W_{\mathrm{ph}}(\hbar\omega) d(\hbar\omega) \ (\text{W/m-K}).} \qquad (9.32)$$

It is useful to write eqn. (9.32) in a more compact form. Let's define the number of modes that participate in transport as

$$\langle M_{\mathrm{ph}}/A \rangle \equiv \int \frac{M_{\mathrm{ph}}(\hbar\omega)}{A} W_{\mathrm{ph}}(\hbar\omega) d(\hbar\omega). \qquad (9.33)$$

Then if we multiply and divide eqn. (9.32) by $\langle M_{\mathrm{ph}} \rangle$, we can express the result as

$$\kappa_L = \frac{\pi^2 k_B^2 T_L}{3h} \langle M_{\mathrm{ph}}/A \rangle \langle\langle \lambda_{\mathrm{ph}} \rangle\rangle, \qquad (9.34)$$

where the average mean-free-path is defined as

$$\langle\langle \lambda_{\mathrm{ph}} \rangle\rangle \equiv \frac{\int \frac{M_{\mathrm{ph}}(\hbar\omega)}{A} \lambda_{\mathrm{ph}}(\hbar\omega) W_{\mathrm{ph}}(\hbar\omega) d(\hbar\omega)}{\int \frac{M_{\mathrm{ph}}(\hbar\omega)}{A} W_{\mathrm{ph}}(\hbar\omega) d(\hbar\omega)}. \qquad (9.35)$$

To summarize, we can write the near-equilibrium lattice heat flux and electrical currents in similar forms

$$\boxed{\begin{aligned} J_Q &= -\kappa_L \frac{dT_L}{dx} \ (\text{W/m}^2) \\ J &= \sigma \frac{d\,(F_n/q)}{dx} \ (\text{A/m}^2) \\ \kappa_L &= \frac{\pi^2 k_B^2 T_L}{3h} \langle M_{\mathrm{ph}}/A \rangle \langle\langle \lambda_{\mathrm{ph}} \rangle\rangle \\ \sigma &= \frac{2q^2}{h} \langle M_{\mathrm{el}}/A \rangle \langle\langle \lambda_{\mathrm{el}} \rangle\rangle, \end{aligned}} \qquad (9.36)$$

Gradients in temperature give rise to heat flow, and gradients in the electrochemical potential give rise to current flow. The thermal conductivity and the electrical conductivity are each a product of their quantum of conductance, the number of modes that participate in transport, and the average mean-free-path for backscattering. To evaluate these expressions, we must specify $M_{ph}(\hbar\omega)$ and $\lambda(\hbar\omega)$. These quantities are discussed in Sec. 9.5 and 9.6.

Exercise 9.1: Relate the lattice thermal conductivity to the lattice specific heat

One often sees elementary derivations of the lattice specific heat with the result [1-4],

$$\boxed{\kappa_L = \frac{1}{3}\left\langle\left\langle\Lambda_{ph}\right\rangle\right\rangle\left\langle v_{ph}\right\rangle C_v \,,} \tag{9.37}$$

where $\left\langle\left\langle\Lambda_{ph}\right\rangle\right\rangle$ is an appropriately-defined mean-free-path, $\left\langle v_{ph}\right\rangle$ is an average phonon velocity, and C_v is the lattice specific heat at constant volume. Derive eqn. (9.37) from the approach in this lecture, and specify how the average mean-free-path and velocity are defined.

i) specific heat

The total energy (per unit volume) of the lattice vibrations is given by

$$E_L = \int_0^\infty (\hbar\omega)\, D_{ph}(\hbar\omega)\, n_0(\hbar\omega) d(\hbar\omega)\,, \tag{9.38}$$

where D_{ph} is the phonon density of states. The *specific heat at constant volume* is the change in energy per degree change in T_L,

$$C_v \equiv \frac{\partial E_L}{\partial T_L}$$

$$= \frac{\partial}{\partial T_L}\int_0^\infty (\hbar\omega)\, D_{ph}(\hbar\omega)\, n_0(\hbar\omega) d(\hbar\omega) \tag{9.39}$$

$$\approx \int_0^\infty (\hbar\omega)\, D_{ph}(\hbar\omega)\left(\frac{\partial n_0(\hbar\omega)}{\partial T_L}\right) d(\hbar\omega)\,.$$

using eqns. (9.17) and (9.25) we find

$$C_v \approx \frac{\pi^2 k_B^2 T_L}{3}\int_0^\infty D_{ph}(\hbar\omega) W_{ph}(\hbar\omega) d(\hbar\omega)\,. \tag{9.40}$$

Next, we multiply and divide eqn. (9.32) by C_v to find

$$\kappa_L = \left\{ \frac{\frac{1}{h} \int \frac{M_{ph}(\hbar\omega)}{A} \lambda_{ph}(\hbar\omega) W_{ph}(\hbar\omega) d(\hbar\omega)}{\int_0^\infty D_{ph}(\hbar\omega) W_{ph}(\hbar\omega) d(\hbar\omega)} \right\} C_v . \tag{9.41}$$

To simplify this expression, we need to examine both λ_{ph} and M_{ph}.

ii) mean-free-path: λ_{ph} vs. Λ_{ph}

Recall from Lecture 6, that the mean-free-path for backscattering in an isotropic 3D material is

$$\lambda_{ph}(\hbar\omega) = \frac{4}{3} \upsilon_{ph}(\hbar\omega) \tau_{ph}(\hbar\omega) . \tag{9.42}$$

Alternatively, the "mean-free-path" is commonly defined as

$$\Lambda_{ph}(\hbar\omega) = \upsilon_{ph}(\hbar\omega) \tau_{ph}(\hbar\omega) , \tag{9.43}$$

so

$$\lambda_{ph}(\hbar\omega) = \frac{4}{3} \Lambda_{ph}(\hbar\omega) . \tag{9.44}$$

iii) $M_{ph}(\hbar\omega)$ and $D_{ph}(\hbar\omega)$

In Lecture 2 we learned that $M(E)$ is related to $D(E)$ by

$$M_{el}(E) = A \frac{h}{4} \langle \upsilon_x^+ \rangle D_{el}(E) , \tag{9.45}$$

where $\langle \upsilon_x^+ \rangle$ is the average velocity in the direction of transport at energy, E. For spherical bands in 3D

$$\langle \upsilon_\tau^+ \rangle = \frac{\upsilon_{el}(E)}{2} . \tag{9.46}$$

It is important to note that the electron density of states contains a factor of 2 for spin. In terms of the density of states per spin, $D'_{el}(E) = D_{el}(E)/2$, eqn. (9.45) becomes

$$M_{el}(E) = A \frac{h}{2} \langle \upsilon_x^+ \rangle D'_{el}(E) , \tag{9.47}$$

Since there is no spin for phonons, we can find M_{ph} directly from eqn. (9.47) to obtain

$$M_{ph}(\hbar\omega) = A \frac{h}{2} \left(\frac{\upsilon_{ph}(\hbar\omega)}{2} \right) D_{ph}(\hbar\omega) = A \frac{h}{4} \upsilon_{ph}(\hbar\omega) D_{ph}(\hbar\omega) , \tag{9.48}$$

Finally, by inserting eqns. (9.44) and (9.48) in (9.41), we find

$$\kappa_L = \left\{ \frac{\frac{1}{3} \int \Lambda(\hbar\omega) v_{\text{ph}}(\hbar\omega) D_{\text{ph}}(\hbar\omega) W_{\text{ph}}(\hbar\omega) d(\hbar\omega)}{\int_0^\infty D_{\text{ph}}(\hbar\omega) W_{\text{ph}}(\hbar\omega) d(\hbar\omega)} \right\} C_v , \qquad (9.49)$$

which is still not in the final form, eqn. (9.37) that we seek.

iv) average mean-free-path and velocity

The next step is to multiply and divide eqn. (9.49) by

$$\int v_{\text{ph}} D_{\text{ph}} W_{\text{ph}} d(\hbar\omega) , \qquad (9.50)$$

after which we can group terms and define an average mean-free-path as

$$\langle\langle \Lambda_{\text{ph}} \rangle\rangle \equiv \frac{\int \Lambda_{\text{ph}}(\hbar\omega) v_{\text{ph}}(\hbar\omega) D_{\text{ph}}(\hbar\omega) W_{\text{ph}}(\hbar\omega) d(\hbar\omega)}{\int v_{\text{ph}}(\hbar\omega) D_{\text{ph}}(\hbar\omega) W_{\text{ph}}(\hbar\omega) d(\hbar\omega)} \qquad (9.51)$$

and an average velocity as

$$\langle v_{\text{ph}} \rangle \equiv \frac{\int v_{\text{ph}}(\hbar\omega) D_{\text{ph}}(\hbar\omega) W_{\text{ph}}(\hbar\omega) d(\hbar\omega)}{\int D_{\text{ph}}(\hbar\omega) W_{\text{ph}}(\hbar\omega) d(\hbar\omega)} . \qquad (9.52)$$

Finally, using eqns. (9.51) and (9.52) in (9.49), we find the desired result, eqn. (9.37).

Equation (9.37) is often used to estimate the average mean-free-path from the measured heat capacity and thermal conductivity. To do this, we need to know the average velocity, which is frequently assumed to be the longitudinal sound velocity. Our derivation has identified the precise definitions of the average mean-free-path and velocity. Given a phonon dispersion, for example, we can compute the average velocity from eqn. (9.49). It is typically very different from the longitudinal sound velocity, which means estimates of the average mean-free-path can be very wrong if they assume the longitudinal sound velocity [5].

9.5   Debye model for $M_{\text{ph}}(\hbar\omega)$

For electrons, the effective mass model is widely-used and generally produces reasonably accurate results. This occurs because, as shown in Fig. 9.3(a), the bandwidth of the electronic dispersion is typically $\gg k_B T_L$, so only states near the bottom of the conduction band where the effective mass model is reasonably accurate are occupied. For phonons, however, the situation is much different. As illustrated in Fig. 9.3(b), the bandwidth of the phonon dispersion is on the order of $k_B T_L$, so states across the entire Brillouin zone are occupied. The widely-used Debye approximation, $\omega = vq$, sketched in Fig. 9.1(b), fits the acoustic branches, as long as q is not too far from the center of the Brillouin zone.

To make use of the Debye approximation, we need to be careful. First, we write

$$\hbar\omega = \hbar v_D q \,, \tag{9.53}$$

where v_D, the *Debye velocity*, is an average velocity of the longitudinal and transverse acoustic modes. Then it is easy to find the density of states just as we do for electron states. The answer is

$$D_{\text{ph}}(\hbar\omega) = \frac{3(\hbar\omega)^2}{2\pi^2(\hbar v_D)^3} \quad (\text{No./J-m}^3)\,. \tag{9.54}$$

The factor of three is for the three polarizations, LA and two TA. A word of caution. We have computed $D_{\text{ph}}(\hbar\omega)$, the density-of-states per unit volume per Joule because we are working in phonon energy space to keep the analogy to the electron density-of-states in energy space. In textbooks, one usually sees, $D_{\text{ph}}(\omega)$ the density-of-states per unit volume per Hz. The two are related by

$$D_{\text{ph}}(\hbar\omega) = \frac{D_{\text{ph}}(\omega)}{\hbar}\,. \tag{9.55}$$

Having computed the density-of-states, we can obtain the number of phonon modes per cross-sectional area from eqn. (9.48) to find

$$M_{\text{ph}}(\hbar\omega) = \frac{3(\hbar\omega)^2}{4\pi(\hbar v_D)^2}\,, \tag{9.56}$$

but this is not enough. Since all of the states in the Brillouin tend to be occupied at moderate temperatures, we need to be sure that we have the correct number of states. For a crystal with N/Ω atoms per unit volume, there are $3N/\Omega$ states per unit volume. To find the total number of states we integrate the density-of-states,

$$\int_0^{\hbar\omega_D} D_{\text{ph}}(\hbar\omega) d(\hbar\omega)\,. \tag{9.57}$$

The upper limit of this integral, $\hbar$ times the so-called *Debye frequency*, is chosen so to produce the correct total number of states, $3N/\Omega$. The result is

$$\hbar\omega_D = \hbar v_D \left(\frac{6\pi^2 N}{\Omega}\right)^{1/3} \equiv k_B T_D\,. \tag{9.58}$$

The Debye frequency defines a cutoff frequency above which no states are permitted. Alternatively, we could express this as a cutoff wavevector, q_D, beyond which no states are permitted. This cutoff can also be expressed as a *Debye temperature*, $T_D = \hbar\omega/k_B$. For $T_L \ll T_D$, only states with small wavevectors for which the Debye approximation is accurate are occupied.

Now returning to eqn. (9.32), we can find the lattice thermal conductivity by integrating to the Debye cutoff energy,

$$\kappa_L = \frac{\pi^2 k_B^2 T_L}{3h} \int_0^{\hbar\omega_D} \frac{M_{\mathrm{ph}}(\hbar\omega)}{A} \lambda_{\mathrm{ph}}(\hbar\omega) W_{\mathrm{ph}}(\hbar\omega) d(\hbar\omega) \ \ (\mathrm{W/m\text{-}K}) \quad (9.59)$$

and using the Debye approximation, eqn. (9.56) for M_{ph}. The integral must be done numerically, but if we can develop expressions for the mean-free-path, then the integral can be done. This is how lattice thermal conductivities were first calculated [7, 8].

With fast digital computers, it is relatively straightforward these days to compute the "exact" $M_{\mathrm{ph}}(\hbar\omega)$ from an accurate phonon dispersion [5]. Figure 9.6 displays some example calculations. Figure 9.6(a) shows that the Debye result for M_{ph} is a good approximation to the exact result only for $\hbar\omega < 0.02$ eV. For $T_L = 50$ K, the window function is restricted to low energies, and only the states for which the Debye approximation holds are occupied. For 300 K, however, the window function is very broad. All of the states are occupied, and the Debye approximation is poor. For silicon, $T_D \approx 640$ K, so T_L must be very much less than this value to ensure the accuracy of the Debye approximation. Figure 9.6(b), a similar comparison for electrons, shows that the effective mass approximation is quite good. At both low and high temperatures, only low energy channels are occupied, and the effective mass approximation is good.

9.6   Phonon scattering

Equations (9.36) show that both the electrical and lattice thermal conductivities are proportional the average mean-free-path for backscattering. Electrons scatter from defects (ionized impurities, neutral impurities, crystal defects, etc.), from phonons, from roughness at surfaces and boundaries, and from other electrons. As discussed in Lecture 6, we compute scattering rates by using Fermi's Golden Rule.

Phonons can also scatter from defects (not charged defects, but impurity atoms, isotopes, etc.) from other phonons, from surfaces and boundaries, and from electrons. Phonon-electron scattering can give rise to *phonon*

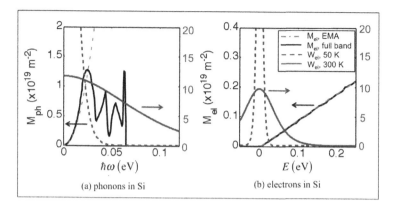

Fig. 9.6. Comparison of the actual distribution of channels in silicon with simple approximations. (a) Exact distribution of phonon channels (solid line) compared with the Debye approximation, eqn. (9.56) (dashed line). Also shown (right axis) are the window functions at 300 K (solid line) and 50 K (dashed line). (The calculations for electrons use the methods described in [6].) (b) Exact distribution of electron channels (solid line) compared with the effective mass approximation, eqn. (5.55) (dashed line on top of the solid line). Also shown are the window functions at 300 K (solid line) and 50 K (dashed line). Phonon results are after Jeong, *et al.* [5].

drag, which can become important at low temperatures. Just as for electrons, we compute phonon scattering rates by using Fermi's Golden Rule.

Phonon-phonon scattering occurs because the potential energy of the bonds in the crystal are not exactly harmonic. The first term in a Taylor series expansion of the potential is harmonic (i.e. $U \propto \delta r^2$, where δr is the deviation from the equilibrium lattice spacing), but higher order terms are treated as a scattering potential. Figure 9.7 illustrates two types of phonon scattering. In the normal (N) process (Fig. 9.7(a)), two phonons interact and create a third phonon. Energy and momentum must be conserved, so

$$\hbar \vec{q_3} = \hbar \vec{q_1} + \hbar \vec{q_2}$$
$$\hbar \omega_3 = \hbar \omega_1 + \hbar \omega_2 . \tag{9.60}$$

This type of scattering has little effect on the heat flux because the total momentum of the phonon ensemble is conserved.

A second type of scattering, Umklapp or U-scattering is illustrated in Fig. 9.7(b). In this case, the two initial phonons have larger momentum, so that when they scatter, the resulting phonon would have a momentum outside the Brillouin zone. This value of the momentum is non-physical because it implies a wavelength that is shorter than the lattice spacing. Instead, we find the final momentum by adding a reciprocal lattice vector,

$\vec{G}$. As shown in Fig. 9.7(b), this type of scattering reverses the x-directed momentum, so it does affect the heat current. It may be easier to understand U-scattering in a repeated zone plot of $\hbar\omega$ vs. q. The final state (the dashed line in Fig. 9.7(b)) lies in a portion of the next Brillouin zone where the velocity is opposite to the incident velocity. For U-scattering to occur, a large population of large q phonons is needed, so the U-processes become important at high temperatures.

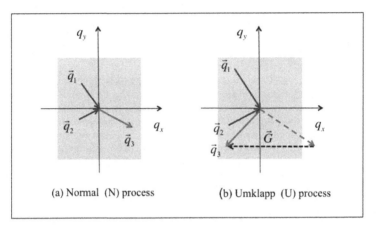

Fig. 9.7.   Illustration of phonon scattering processes. (a) Normal or N-processes, which conserve phonon crystal momentum and (b) Umklapp or U-processes, which do not conserve crystal momentum.

Scattering rates add, so the total phonon scattering rate is

$$\frac{1}{\tau_{\mathrm{ph}}(\hbar\omega)} = \frac{1}{\tau_D(\hbar\omega)} + \frac{1}{\tau_B(\hbar\omega)} + \frac{1}{\tau_U(\hbar\omega)}, \qquad (9.61)$$

where the first term describes scattering from defects, the second scattering from boundaries, and the third phonon-phonon scattering by U processes. Alternatively, we could write eqn. (9.61) in terms of the mean-free-paths,

$$\frac{1}{\lambda_{\mathrm{ph}}(\hbar\omega)} = \frac{1}{\lambda_D(\hbar\omega)} + \frac{1}{\lambda_B(\hbar\omega)} + \frac{1}{\lambda_U(\hbar\omega)}, \qquad (9.62)$$

where for 3D spherical systems, $\lambda = (4/3)v_{\mathrm{ph}}\tau_{\mathrm{ph}}$.

Expressions for each of the scattering rates can be developed. For example, for scattering from point defects, we find [4]

$$\frac{1}{\tau_D(\hbar\omega)} \propto \omega^4. \qquad (9.63)$$

This type of scattering is known a *Rayleigh scattering*; it is like the scattering of light from dust particles in the atmosphere. Higher frequency (short wavelength) phonons "feel" these small defects more and are, therefore, more strongly scattered.

For boundaries and surfaces, we can write [4]

$$\frac{1}{\tau_B(\hbar\omega)} \propto v_{ph}(\hbar\omega)/L, \qquad (9.64)$$

where L is the shortest dimension of the sample. This type of scattering is inversely proportional to the time it takes for a phonon to travel to the boundary.

The third important process, Umklapp scattering, is harder to describe simply. A commonly used expression is [4]

$$\frac{1}{\tau_U(\hbar\omega)} \propto e^{-T_D/bT_L}T_L^3\omega^2. \qquad (9.65)$$

With this background, we are now able to understand the temperature-dependent lattice thermal conductivity.

9.7 Discussion

i) lattice thermal conductivity vs. temperature

Figure 9.8 is a plot of the measured lattice thermal conductivity of silicon vs. temperature. The solid line is a theoretical calculation using techniques discussed in this lecture and in [5]. According to eqn. (9.36), κ_L is proportional to the number of channels that are occupied, $\langle M_{ph}\rangle$, and to the average mean-free-path, $\langle\langle\lambda_{ph}\rangle\rangle$. The characteristic in Fig. 9.7 can be understood by understanding how $\langle M_{ph}\rangle$ and $\langle\langle\lambda_{ph}\rangle\rangle$ vary with temperature.

Using eqn. (9.33), it can be shown that at low temperatures, $\langle M_{ph}\rangle$ varies as T_L^3 (the same is true of the specific heat). The initial rise in thermal conductivity is due to the fact that the number of populated channels rises quickly with temperature. At low temperatures, boundary scattering is important. As the temperature increases, more high q (short wavelength) phonons are produced. These shorter wavelength phonons "see" point defects, so defect scattering becomes important. As the temperature approaches the Debye temperature, all of the phonon modes are populated, and further increases in temperature do not change $\langle M_{ph}\rangle$. Instead, the high temeperatures lead to increased phonon scattering by U-processes, so the thermal conductivity drops with increasing temperatures.

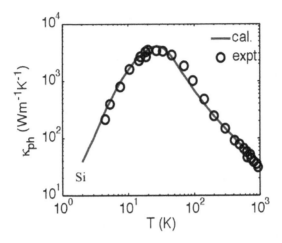

Fig. 9.8.　The measured and simulated thermal conductivity of bulk silicon as a function of temperature. (The calculated results use the methods of Jeong, *et al.* [5], and the data points are from C.J. Glassbrenner and G.A. Slack, "Thermal Conductivity of Silicon and Germanium from 3 K to the Melting Point", *Phys. Rev.*, **134**, A1058, 1964.)

ii) difference between lattice thermal and electrical conductivities

The expressions for the lattice thermal conductivity and the electrical conductivity as given by eqns. (9.36) are very similar. In practice, the average electron and phonon mean-free-paths are of the same order of magnitude. Why then does the electrical conductance vary over many more orders of magnitude while the lattice thermal conductivity only varies over a few? The answer lies in the window functions as given by eqns. (9.22) and (9.25). For both electrons and phonons, higher temperatures broaden the window function and increase the population of states. Electrons are fermions, however, and the position of the Fermi level has a dramatic effect on the magnitude of the window function. By controlling the position of the Fermi level, the electrical conductivity can be varied over many orders of magnitude.

Another key difference between electrons and phonons comes from how the states are populated. For a thermoelectric device, $E_F \approx E_c$, and the electron and phonon window functions are quite similar. For electrons, however, the bandwidth of the dispersion is very large, so only a few states near the bottom of the band are populated. The effective mass approximation works well for these states, so it is easy to get analytical solutions. For phonons, the bandwidth of the dispersion is small. At moderate temperatures, states all across the entire Brillouin zone are occupied. Simple, analytical approximations do not work well, so it is hard to obtain analytical solutions for the lattice thermal conductivity.

iii) quantized lattice thermal conductivity

In Lecture 3 Sec. 3.2, we discussed quantized electrical conductance. Consider eqn. (9.36) at low temeperatures where $W_{\mathrm{ph}}(\hbar\omega)$ is sharply peaked near $\hbar\omega = 0$. Accordingly, eqn. (9.36) becomes

$$K_L = \frac{\pi^2 k_B^2 T_L}{3h} T_{\mathrm{ph}}(0) M_{\mathrm{ph}}(0) \,. \tag{9.66}$$

For a bulk material, $M_{\mathrm{ph}}(\hbar\omega) \to 0$ as $\hbar\omega \to 0$, but for a nanostructure, such as a nanowire or nanoribbon, one can have a finite number of phonon modes. For ballistic phonon transport, $T_{\mathrm{ph}} = 1$, so we expect to see $K_L = \pi^2 k_B^2 T_L / 3h$ times the number of zero frequency modes. In an experiment with 4 modes, just such a result was observed [9]. Given the close analogy between electron and phonon transport, quantized heat flow is expected. The experiments are easier to do with electrons because with electrical gates it is possible to control the dimensions of a channel and, therefore, the number of modes. For phonons, the number of modes is fixed by the physical structure.

9.8 Summary

This lecture has been a very short introduction to phonon transport. We saw that the concepts used to describe electron transport can be generalized for phonons. The Landauer approach for phonon transport describes transport all the way from ballistic to diffusive regimes. Thermal transport is even quantized, just as electron transport is, and quantized thermal transport has been observed in nanostructures. In the diffusive regime, the results are equivalent to well-known, traditional approaches based, for example, on the phonon Boltzmann Transport Equation. We saw that the lattice thermal conductivity can be written in a form that is very similar to the electrical conductivity, but there are two important differences.

The first difference between electrons and phonons is the difference in bandwidths of their dispersions. The bandwidth of the phonon dispersion is on the order of $k_B T_L$, so at room temperature, all of the acoustic modes across the entire Brillouin zone can be occupied. For electrons, the dispersion typically has a bandwidth much greater than $k_B T_L$ at room temperature, so only low energy, small wavevector electron states are occupied. As a result, the simple Debye approximation to the acoustic phonon dispersion does not work nearly as well as the simple effective mass approximation to the electron dispersion.

The second difference between electrons and phonons is that for electrons the population of the modes is controlled by the window function, which depends on the position of the Fermi level and the temperature. For phonons, the window function depends only on the temperature. The results is that electrical conductivities vary over many orders of magnitude as the position of the Fermi level varies, while lattice conductivities vary over only a few orders of magnitude.

This lecture is only a starting point — much more could be said about thermal conduction. For example, the engineering of lattice thermal conductivities by nanostructuring materials is a current research topic of great interest.

9.9   References

For a thorough introduction to phonons, see these classic texts:

[1] J.M. Ziman, *Principles of the Theory of Solids*, Cambridge Univ. Press, Cambridge, U.K., 1964.

[2] N.W. Ashcroft and N.D. Mermin, *Solid–State Physics*, Saunders College, Philadelphia, PA, 1976.

[3] C. Kittel, *Introduction to Solid State Physics*, 4th Ed., John Wiley and Sons, New York, 1971.

For a recent treatment that covers electrons and phonons and thermo-electeric effects, see

[4] Gang Chen, *Nanoscale Energy Transport and Conversion*, Oxford Univ. Press, New York, 2005.

The application of the techniques discussed in this chapter to real materials is discussed in:

[5] C. Jeong, S. Datta, M. Lundstrom, "Full Dispersion vs. Debye Model Evaluation of Lattice Thermal Conductivity with a Landauer approach", *J. Appl. Phys.*, **109**, 073718-8, 2011.

The close relation between the approach for phonons discussed in this lecture and that for electrons is apparent by comparing the following paper by Jeong, et al. to the previous paper.

[6] Changwook Jeong, Raseong Kim, Mathieu Luisier, Supriyo Datta, and Mark Lundstrom, "On Landauer vs. Boltzmann and Full Band vs. Effective Mass Evaluation of Thermoelectric Transport Coefficients", *J. Appl. Phys.*, **107**, 023707, 2010.

Two of the early and still highly cited papers on lattice heat conduction are by Callaway and Holland.

[7] J. Callaway, "Model for lattice thermal conductivity at low temperatures", *Phys. Rev.*, **113**, 1046-1051, 1959.

[8] M.G. Holland, "Analysis of lattice thermal conductivity", *Phys. Rev.*, **132**, 2461-2471, 1963.

Quantized thermal conductance was first observed by Schwab, et al.

[9] K. Schwab, E. A. Henriksen, J. M. Worlock, and M. L. Roukes, "Measurement of the quantum of thermal conductance", *Nature*, **404**, 974-977, 2000.

Lecture 10

Graphene: A Case Study

Contents

10.1 Introduction

In these lectures we have developed an understanding of near-equilibrium transport along with the mathematical theory to compute transport coefficients. The coupled current equations can be written as in eqns. (5.26):

$$J_x = \sigma \frac{d\left(F_n/q\right)}{dx} - S\sigma \frac{dT_L}{dx}$$
$$J_{Qx} = \pi\sigma \frac{d\left(F_n/q\right)}{dx} - \kappa_0 \frac{dT_L}{dx} \,,$$

$$(10.1)$$

or, in the inverted form as in eqns. (5.30):

$$\frac{d\left(F_n/q\right)}{dx} = \rho J_x + S\frac{dT_L}{dx}$$
$$J_{Qx} = \pi J_x - \kappa_e \frac{dT_L}{dx} \,.$$

$$(10.2)$$

The transport coefficients in 2D are:

$$\sigma_S = 1/\rho_S = \int \sigma'(E)dE$$

$$S = -\left(\frac{k_B}{q}\right) \int \left(\frac{E - E_F}{k_B T_L}\right) \sigma'(E)dE/\sigma$$

$$\pi = T_L S$$

$$\kappa_0 = T_L \left(\frac{k_B}{q}\right)^2 \int \left(\frac{E - E_F}{k_B T_L}\right)^2 \sigma'(E)dE$$

$$\kappa_e = \kappa_0 - \pi S \sigma$$

$$\sigma'(E) = \frac{2q^2}{h} \frac{M(E)}{W} \lambda(E) \left(-\frac{\partial f_0}{\partial E}\right) .$$

(10.3)

Equations (10.1)–(10.3) are valid for diffusive transport in either n- or p-type materials and also when conduction is through both the conduction and valence bands. A two-dimensional conductor has been assumed; for 3D, replace $M(E)/W$ by $M(E)/A$ and for 1D by $M(E)$. To treat transport from the diffusive to the ballistic limits, replace the mean-free-path by the apparent mean-free-path, $1/\lambda_{\mathrm{app}} = 1/\lambda + 1/L$. To evaluate these equations, we only need to specify two parameters, $M(E)$ and $\lambda(E)$. For parabolic energy bands, $M(E)$ is given by eqns. (2.31), but eqns. (10.1)–(10.3) are valid for any band structure; we just need to use the appropriate $M(E)$. The mean-free-path, $\lambda(E)$ also needs to be specified according the the physics of the scattering processes.

Graphene provides us with a good case study on applying the approaches described in these lectures to new materials. The band structure is simple, but distinctly non-parabolic. It is a material of great scientific interest, as recognized by the 2010 Noble Prize in physics [1], and there is also much interest in potential technological applications. For our purposes, however, graphene simply provides us with an example of how to compute transport coefficients when the band structure is non-parabolic.

10.2   Graphene

Graphene is a one atom thick, planar sheet of carbon atoms arranged in a honeycomb lattice. Techniques to produce large areas of high quality graphene have recently been developed. The striking feature of graphene is its unusual band structure. As shown in Fig. 10.1, the conduction and valence bands meet at the six vertices of a 2D Brillouin zone. There is no bandgap. Under charge neutral conditions, the Fermi level lies at the intersection of the two bands. As shown in Fig. 10.1(b), only two of the six points in the Brillouin zone, are distinct, the others can be reached by a reciprocal lattice vector. Consequently, the valley degeneracy for graphene is $g_v = 2$. For a tutorial on graphene band structure, see Datta [2].

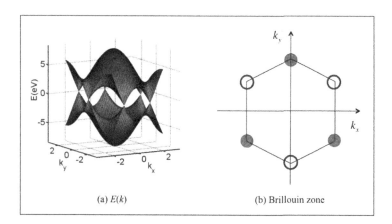

(a) $E(k)$ (b) Brillouin zone

Fig. 10.1. The band structure of graphene as computed from a simple tight-binding model. (a) $E(k)$ and (b) the Brillouin zone showing the six k-points where the conduction and valence bands meet, two of which are distinct.

For near-equilibrium transport, only states near the Fermi level where the conduction and valence bands of graphene intersect are important. We can, therefore, simplify the band structure as shown in Fig. 10.2. The conduction and valence bands are two cones that meet at $E = 0$, which is called the *neutral point* or *Dirac point*. Near the Dirac point, we have

$$E(k) = \pm \hbar v_F k = \pm \hbar v_F \sqrt{k_x^2 + k_y^2}. \tag{10.4}$$

We will refer to the case where $E_F > 0$ as n-type graphene, and $E_F < 0$ as p-type graphene. Note that the electron velocity is constant, independent of k,

$$v(k) = \frac{1}{\hbar} \frac{\partial E}{\partial k} \equiv v_F \approx 1 \times 10^8 \, \text{cm/s}. \tag{10.5}$$

The linear dispersion of graphene near the Dirac point gives a constant velocity. For graphene, this value is very high, about 10^8cm/s, which has led to considerable interest for possible applications in high-speed electronics.

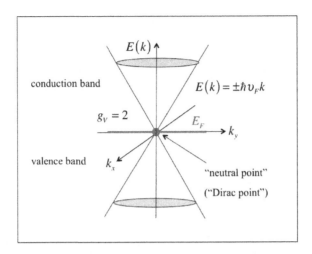

Fig. 10.2. Simplified band structure of graphene for energies near the Dirac point.

We should also mention that the electron wavefunction for graphene
has a special character that leads to some unusual properties. We often
describe electrons in the conduction band of a semiconductor by a simple
envelope function,

$$\psi(x, y) = \frac{1}{\sqrt{A}} e^{i(k_x x + k_y y)} , \tag{10.6}$$

where A is a normalization area. (The actual wavefunction is a product
of the envelope function and a Bloch function with the periodicity of the
lattice.) There are valence bands too. For common, cubic semiconductors,
we have light hole, heavy hole, and split off bands. In principle, we should
deal with a wavefunction that couples all of these bands, but if the bandgap
is large, we can deal with the conduction and valence bands separately. For
graphene, there is no bandgap, and the conduction and valence bands are
always coupled. As a result, electrons in graphene are described by a *two-component wavefunction*,

$$\begin{pmatrix} \psi_a \\ \psi_b \end{pmatrix} = \frac{1}{\sqrt{2A}} \begin{pmatrix} 1 \\ s e^{i\theta} \end{pmatrix} e^{i(k_x x + k_y y)} , \tag{10.7}$$

$$s = \text{sgn}(E)$$
$$\theta = \arctan(k_y / k_x) . \tag{10.8}$$

This wavefunction has some interesting consequences. For example, electrons along $+k_x$ and $-k_x$ have orthogonal wavefunctions, so there is no
probability of backscattering by 180 degrees.

10.3   Density-of-states and carrier density

Our aim is to compute the conductivity of graphene, but the conductivity depends on E_F, and the location of E_F depends on the carrier density. Experimentally, the location of the Fermi level is controlled by controlling the carrier density. To compute the carrier density, we need the density-of-states.

Calculation of the density-of-states for graphene proceeds as illustrated in Fig. 10.3. (For a review of computing the density-of-states, see [3].) The number of states between k and $k + dk$ is

$$N(k)dk = \frac{2\pi k dk}{(2\pi/L_x)(2\pi/L_y)} \times 2 \times g_v \,, \qquad (10.9)$$

where $L_x L_y = A$ is the area for which the number of states is being computed, the denominator is the k-space area occupied by a 2D k-state, the factor of 2 is for spin, and g_v is the valley degeneracy. Using the dispersion, $E = \hbar v_F k$, we change variables to energy, use $g_v = 2$, and find

$$N(k)dk = A g_v \frac{E dE}{\pi (\hbar v_F)^2} \,. \qquad (10.10)$$

Finally, writing $D(E)dE$ as the number of states per unit area between E and $E + dE$ and recognizing that the energy can be greater or less than zero, we find the density-of-states for graphene to be

$$\boxed{D(E) = \frac{2|E|}{\pi \hbar^2 v_F^2} \,.} \qquad (10.11)$$

The number of electrons in the conduction band is obtained from

$$n_S(E_F) = \int_0^\infty D(E) f_0(E) \, dE \,. \qquad (10.12)$$

Because graphene is degenerate, the $T_L = 0$ K assumption is a good approximation, even at room temperature. Accordingly, eqn. (10.12) becomes

$$n_S(E_F) = \int_0^{E_F} D(E)dE = \frac{2}{\pi \hbar^2 v_F^2} \int_0^{E_F} E dE \,, \qquad (10.13)$$

from which we find

$$\boxed{n_S(E_F) = \frac{E_F^2}{\pi \hbar^2 v_F^2} \,.} \qquad (10.14)$$

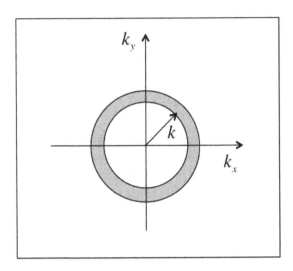

Fig. 10.3. The number of states between k and $k+dk$ is the shaded area, $2\pi k dk$, divided by the area associated with a k-state in two dimensions.

10.4 Number of modes and conductance

According to eqn. (2.25), the number of modes at energy, E, is

$$M(E) = W\frac{h}{4}\left\langle v_x^+(E)\right\rangle D_{2D}(E).$$ (10.15)

For graphene

$$\left\langle v_x^+(E)\right\rangle = \frac{2}{\pi}v_F,$$ (10.16)

and the density-of-states is given by eqn. (10.11), so we find

$$M(E) = W\frac{2|E|}{\pi\hbar v_F}.$$ (10.17)

Figure 10.4 compares $D(E)$ and $M(E)$ for graphene. Note that in contrast to the case for parabolic energy bands where $D(E)$ and $M(E)$ depend differently on energy (Fig. 2.3), for graphene both the density-of-states and number of modes are proportional to energy.

Computing the conductivity is now straight-forward. From eqn. (10.3), we find

$$\begin{aligned}\sigma'(E) &= \frac{2q^2}{h}\frac{M(E)}{W}\lambda(E)\left(-\frac{\partial f_0}{\partial E}\right)\\ &\approx \frac{2q^2}{h}\frac{M(E)}{W}\lambda(E)\,\delta(E-E_F),\end{aligned}$$ (10.18)

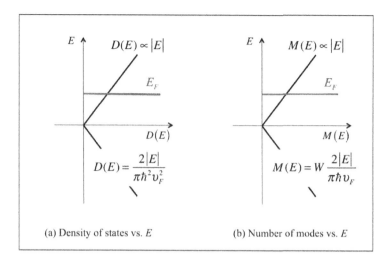

Fig. 10.4. Comparison of the density-of-states and number of modes vs. energy for graphene.

where we have again made the $T_L = 0$ K assumption. The sheet conductance is

$$\sigma_S = \int \sigma'_n(E)dE = \frac{2q^2}{h}\frac{M(E_F)}{W}\lambda(E_F)$$
$$= \frac{2q^2}{h}\left(\frac{2E_F}{\pi\hbar v_F}\right)\lambda(E_F). \tag{10.19}$$

We have derived a simple expression for the sheet conductance of graphene. When the mean-free-path is independent of energy, the conductivity is proportional to E_F, which means that it is proportional to $\sqrt{n_S}$.

10.5 Scattering

The shape of the conductance vs. E_F or conductance vs. n_S characteristic is determined by how the mean-free-path depends on energy. Recall that the mean-free-path for backscattering is given by

$$\lambda(E) = \frac{\pi}{2}v_F\tau_m(E), \tag{10.20}$$

so to understand $\lambda(E)$, we need to understand how the momentum relaxation time, $\tau_m(E)$ varies with energy.

We learned in Lecture 6 that for short range scattering potentials and for acoustic phonon scattering, the scattering rate is proportional to the density-of-states,

$$\frac{1}{\tau(E)} = \frac{1}{\tau_m} \propto D(E) \propto E.$$ (10.21)

Accordingly, $\tau_m \propto E^{-1}$ and

$$\lambda(E) \propto E^{-1}.$$ (10.22)

What does this type of scattering mean for the sheet conductance? According to eqn. (10.19) we must conclude that σ_S is constant — independent of E_F or n_S. Since $\sigma_S = n_S q \mu_n$, we also conclude that the mobility is inversely proportional to n_S. This is an unusual situation. Normally, the higher the carrier density the higher the conductivity. When this type of scattering dominates, the conductivity of graphene is independent of the carrier density.

As discussed in Chapter 6, ionized impurity scattering introduces fluctuations in the potential that scatter carriers. Higher energy carriers "see" these fluctuations less and are consequently scattered less. The mean-free-path increases with energy. Calculations show that the mean-free-path varies linearly with energy [4],

$$\lambda(E) = \lambda_{\mathrm{II}} E,$$ (10.23)

which, when inserted in eqn. (10.19), gives a conductivity that goes as E_F^2. The carrier density, n_S, also varies as E_F^2, so we conclude that charged impurity scattering leads to a conductivity that varies as n_S. The result from $\sigma_S = n_S q \mu_n$ shows that charged impurity scattering leads to a constant mobility.

Several other scattering mechanisms can be important in graphene. These include short range scattering due to defects in the honeycomb lattice and polar phonons in the SiO_2 that the graphene often lies on. For a comprehensive review of the electronic properties of graphene, see [4]. The general features of the conductivity vs. gate voltage characteristic are readily understood as illustrated in Fig. 10.5. Two scattering mechanisms are assumed: 1) ionized impurity scattering which produces a linear $\sigma_S(n_S)$ vs. n_s characteristic and 2) ADP or short range scattering, which produces a constant $\sigma_S(n_S)$ vs. n_S characteristic. Scattering rates add, so the total mean-free-path is

$$\frac{1}{\lambda_{\mathrm{tot}}} = \frac{1}{\lambda_{\mathrm{II}}} + \frac{1}{\lambda_{SR}},$$ (10.24)

and

$$\frac{1}{\sigma_{\mathrm{tot}}} \approx \frac{1}{\sigma_{\mathrm{II}}} + \frac{1}{\sigma_{SR}}.$$ (10.25)

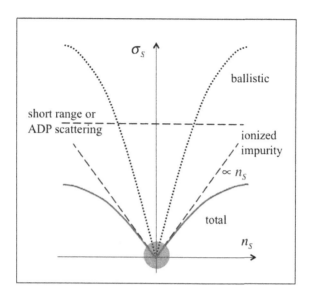

Fig. 10.5. Illustration of the expected shape of the conductivity vs. carrier density characteristic for graphene. Two scattering mechanisms are assumed: 1) ionized impurity scattering and 2) ADP or short range scattering. Also shown is the expected shape of the ballistic characteristic.

At any given carrier density, the smaller of the two contributions limits the total σ_S, so the result is the non-linear characteristic sketched in Fig. 10.5.

Exercise 10.1: Maximum conductivity of graphene

Consider pure graphene with no defects or ionized impurities to scatter electrons. What is the conductivity? At higher temperatures, other phonon-assisted scattering processes occur, but near room temperature, acoustic deformation potential scattering dominates. The scattering rate is proportional to $D(E)$, and the scattering time can be evaluated as [5]:

$$\tau_m(E) = \frac{4\hbar^3 \rho_m \upsilon_F^2 \upsilon_S^2}{D_A^2 k_B T_L} \left(\frac{1}{E}\right), \tag{10.26}$$

where ρ_m is the mass density, υ_S the sound velocity, and D_A the acoustic deformation potential, which is a measure of the strength of the electron-phonon coupling. From eqn. (10.20) with (10.26), we find the mean-free-path as

$$\lambda(E) = \frac{2\pi\hbar^3 \rho_m v_F^3 v_S^2}{D_A^2 k_B T_L} \left(\frac{1}{E}\right), \qquad (10.27)$$

and, from eqn. (10.19), a sheet conductance of

$$\sigma_S = \frac{4q^2 \hbar \rho_m v_F^2 v_S^2}{\pi D_A^2 k_B T_L}. \qquad (10.28)$$

Inserting the appropriate values for graphene

$$v_S \approx 2.1 \times 10^4 \, \text{m/s}$$

$$\rho_m \approx 7.6 \times 10^{-7} \, \text{kg/m}^2$$

$$D_A \approx 18 \, \text{eV},$$

we find

$$\rho_S = \frac{1}{\sigma_S} \approx 30 \, \Omega/\square. \qquad (10.29)$$

(As discussed in reference to eqn. (8.3), recall that $\square$ is not a real unit of measurement. The sheet resistance is commonly written as "Ohms per square" because the resistance of a square, $L = W$, resistor is just ρ_S.)

Equation (10.29) gives the upper limit of the conductivity (lower limit sheet resistance) that we should expect. In practice, other scattering mechanisms, would raise the sheet resistance. How does this lower limit sheet resistance compare to other materials? Consider a Si MOSFET with $n_S \approx 10^{13} \, \text{cm}^{-3}$ and an inversion layer mobility of $\mu_{\text{eff}} \approx 250 \, \text{cm}^2/\text{V-s}$. For such a 2D channel, we find $\rho_S \approx 2500 \, \Omega/\square$. For a high mobility channel like InGaAs with $n_S \approx 2 \times 10^{12} \, \text{cm}^{-3}$ and a mobility of $\mu_{\text{eff}} \approx 1 \times 10^5 \, \text{cm}^2/\text{V-s}$, we find $\rho_S \approx 300 \, \Omega/\square$. We conclude that the intrinsic conductivity of graphene is very high.

10.6   Conductance vs. carrier density

Measurements of the near-equilibrium conductivity vs. the location of the Fermi level or carrier density are commonly used to characterize the quality of graphene layers. Experiments are typically done as sketched in Fig. 10.6. A layer of graphene is placed on a layer of SiO_2, which is on a doped silicon substrate. By changing the potential of the Si substrate (the "back gate"), the potential in the graphene can be modulated to vary E_F and, therefore, n_S. Typical oxide thicknesses are 90 or 300 Angstroms. When a single monolayer is placed on an oxide layer with these thicknesses, the change

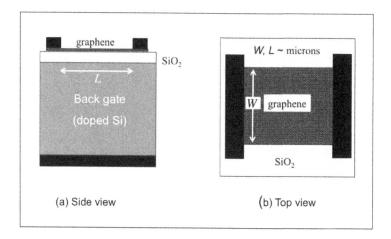

Fig. 10.6. Illustration of the commonly used "back-gating" geometry to characterize the graphene conductivity vs. carrier density. Instead of the two-probe measurement geometry sketched here, four-probe geometries can also be used to eliminate the influence of the contacts.

in color makes single monolayers visible. The location of the Fermi level is fixed by the workfunction of the metal contacts, and the back gate potential moves the Dirac (or neutral) point up and down so that both n- and p-type conduction can be explored. According to simple MOS electrostatics, we expect

$$qn_S \approx C_{\text{ins}} \left(V_G - V_{NP} \right), \tag{10.30}$$

where

$$C_{\text{ins}} = \frac{\epsilon_{\text{ins}}}{t_{\text{ins}}}, \tag{10.31}$$

and V_{NP} is the gate voltage that locates E_F at the Dirac point. The value of V_{NP} is determined by the difference in the workfunction of the metal gate and the graphene and by stray charges. For very thin SiO_2 layers, such as those used in devices, the capacitance of the graphene must also be included, which results in a smaller n_S at a given V_G. (See [5] for a discussion of this effect.)

Figure 10.7 shows the results of typical measurements. In these experiments, pristine graphene was first characterized, and then the sample was exposed to potassium for various times. Several things should be noted. First, note that σ_S does not go to zero at the Dirac point. This occurs because the random distribution of background charges in the system makes

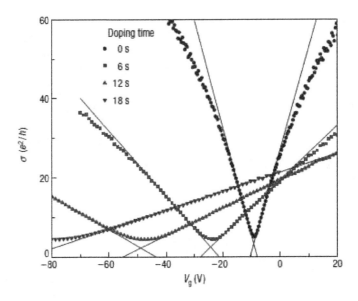

Fig. 10.7.   Measured conductance vs. gate voltage for graphene on SiO$_2$. The conductivity vs. gate voltage characteristic, σ_S vs. V_G, was measured before exposure to potassium and after exposures of various times. (Reprinted by permission from Macmillan Publishers Ltd: *Nature Phys.*, J.-H. Chen, C. Jang, S. Adam, M. S. Fuhrer, E. D. Williams, and M. Ishigami, "Charged-impurity scattering in graphene", **4**, 377-381, copyright 2008.)

it impossible for a single gate voltage to align the Fermi level to the Dirac point everywhere [4]. These background charges (and workfunction differences) also shift the location of the neutral point so that it does not occur at $V_G = 0$ V.

Before exposure to potassium, the results in Fig. 10.7 show a nonlinear σ_S vs. V_G characteristic, reminiscent of the expected characteristic sketched in Fig. 10.5. Increasing exposure to potassium shifts that location of the neutral point to increasingly negative voltages, because of the effect of the charge on the gate electrostatics. It also makes the characteristics increasingly linear, which indicates the increasing dominance of charge impurity scattering. Increasing exposure to potassium also lowers the slope of the σ_S vs. V_G characteristic, which indicates a lowering of the mobility.

The first question to ask is: "How close to the ballistic limit are these results?" From eqns. (10.14) and (10.19), we find

$$\frac{\sigma_S}{q^2/h} = 4\sqrt{\frac{n_S}{\pi}} \lambda_{\text{app}}(E_F). \tag{10.32}$$

For a ballistic sample, $\lambda_{\text{app}} \to L$, where L is the length of the resistor. In this case, $L \approx 10\,\mu\text{m}$. For the pristine sample, the maximum deviation in gate voltage from the neutral point voltage is $\Delta V_{\text{max}} \approx 30$ V. From the insulator thickness ($t_{\text{ins}} = 300$ nm and eqns. (10.30) and (10.31)), we find $n_S|_{\text{max}} \approx 2 \times 10^{12}\,\text{cm}^{-2}$. Inserting these numbers in eqn. (10.32), we find

$$\frac{\sigma_S|_{\text{ball}}}{q^2/h} \approx 3200\,, \qquad (10.33)$$

so the results in Fig. 10.7 are far below the ballistic limit.

The next question to ask is: "How close to the upper (acoustic deformation potential scattering) limit are these results?" In Exercise 10.1, we estimated the limit to be $\rho_S \approx 30\,\Omega/\square$, which translates to $\sigma_S/(q^2/h) \approx 865$, well above the maximum sheet conductances displayed in Fig. 10.7. The nonlinearity observed for the pristine sample is not, therefore, due to the ADP limit, but possibly to short range scattering due to defects. (Such nonlinearities are also common in two probe measurements where they reflect the properties of the contacts.)

Finally, it is of interest to determine the carrier mobility. When the σ_S vs. V_G characteristic is linear, it is straight forward to show that

$$\mu_n = \frac{1}{C_{\text{ins}}} \frac{d\sigma_S}{dV_G}\,. \qquad (10.34)$$

Inserting numbers for the sample with a six second exposure to potassium, we find $\mu_n \approx 3000\,\text{cm}^2/\text{V-s}$, a fairly high mobility for an intentionally doped material, which explains much of the interest in graphene for electronic applications.

Analysis of the gated conductance of graphene has shed much light on the physics of this interesting material. For a review of this work, see [4], and for a tutorial treatment, an online lecture is available [6].

10.7 Discussion

We have shown in this lecture how to apply the concepts developed in previous lectures to new and novel materials, such as graphene. In this section, we briefly discuss a few additional topics.

Mobility and the Drude formula

According to eqn. (10.19),

$$\sigma_S = \frac{2q^2}{h} \frac{M(E_F)}{W} \lambda(E_F) = \frac{2q^2}{h} \left(\frac{2E_F}{\pi \hbar v_F} \right) \lambda(E_F). \qquad (10.35)$$

If we wish to define a mobility, we would do so by equating this result to $\sigma_S = n_S q \mu_n$ to find

$$\mu_n = \frac{2q}{h} \frac{1}{n_S} \left(\frac{2E_F}{\pi \hbar v_F} \right) \lambda(E_F), \qquad (10.36)$$

which, using eqn. (10.14) for n_S can be written as

$$\mu_n = \frac{2q v_F}{\pi E_F} \lambda(E_F). \qquad (10.37)$$

One often hears that electrons in graphene are "massless", which makes sense because the linear dispersion is like the ω vs. k dispersion for photons. Accordingly, to find the mobility, we should not begin with the Drude expression,

$$\mu_n = \frac{q \tau_m}{m^*}, \qquad (10.38)$$

but eqn. (10.37) shows that we can define a mobility without bringing in an effective mass. Nevertheless, a connection can be made. Using eqn. (10.20) to express the mobility in eqn. (10.37) in terms of the scattering time rather than the mean-free-path, we find

$$\mu_n = \frac{q \tau_m(E_F)}{E_F / v_F^2}, \qquad (10.39)$$

which is just like the Drude expression, eqn. (10.38) if we define the effective mass as

$$m^* \equiv \frac{E_F}{v_F^2}. \qquad (10.40)$$

This definition of mass reminds us of the relativistic expression, $E = mc^2$.

Cyclotron resonance frequency

In Chapter 8, we discussed the cyclotron resonance frequency, ω_c, the frequency at which an electron orbits a magnetic field. Equation (8.55) gave

$$\omega_c = \frac{q v B_z}{\hbar k}. \qquad (10.41)$$

For a parabolic energy band, we found

$$\omega_c = \frac{qB_z}{m^*} \tag{10.42}$$

and for graphene we use $v = v_F$ and $E = \hbar v_F k$ to find

$$\omega_c = \frac{qB_z}{E_F/v_F^2} \tag{10.43}$$

so the effective mass for the Drude mobility, eqn. (10.40), works for the cyclotron frequency of graphene too.

Effective mass of graphene

The effective mass form of the mobility and cyclotron frequency suggest that there might be a more general way to define effective mass, rather than the conventional definition that relates it to the curvature of the energy band. Indeed this is a case. Datta shows that a conductivity effective mass can be defined for general (but isotropic) energy bands as [7]

$$m^* \equiv \frac{p}{v} = \frac{\hbar k}{v} \, . \tag{10.44}$$

Equation (10.44) gives the expected result for parabolic energy bands, and it gives eqn. (10.40) for graphene. See Datta [7] for a discussion of this point.

Hall effect for graphene

In Sec. 7.5, we worked out the near-equilibrium magnetoconductivity tensor for low B-fields. For graphene, we begin, again, at eqn. (7.52), but eqn. (7.57) assumes parabolic energy bands. Beginning from eqn. (7.55), which is correct for graphene, the additional term in the BTE due to the B-field is

$$\nabla_p(\delta f) = \nabla_p \left[\tau_m(E) \left(-\frac{\partial f_0}{\partial E} \right) \{ \vec{v} \cdot \vec{G} \} \right]$$

$$= \tau_m(E) \left(-\frac{\partial f_0}{\partial E} \right) \nabla_p \{ \vec{v} \cdot \vec{G} \} \, . \tag{10.45}$$

To evaluate

$$\nabla_p \{ \vec{v} \cdot \vec{G} \} = \frac{\partial}{\partial p_x} (v_x G_x + v_y G_y) \, \hat{x} + \frac{\partial}{\partial p_y} (v_x G_x + v_y G_y) \, \hat{y}, \tag{10.46}$$

we must develop expressions for v_x and v_y. In graphene, the magnitude of the velocity is v_F, a constant, independent of energy, so

$$v_x = v_F \cos\theta = v_F \frac{k_x}{k_x^2 + k_y^2} = \frac{\hbar k_x}{\pm E/v_F^2}, \tag{10.47}$$

where the last result comes from the graphene band structure, eqn. (10.4). A similar result is obtained for v_y, so we conclude that

$$v_x = \frac{p_x}{m^*(E)}$$

$$v_y = \frac{p_y}{m^*(E)}, \tag{10.48}$$

where

$$m^*(E) = +E/v_F^2 \quad (E > 0)$$

$$m^*(E) = -E/v_F^2 \quad (E < 0). \tag{10.49}$$

We see once again the graphene effective mass as defined in eqn. (10.40), but for energies above the Dirac point, the effective mass is positive, and for energies below the Dirac point it is negative.

Now if we return to eqn. (10.46) and use eqns. (10.48), we find

$$\nabla_p \left\{ \vec{v} \cdot \vec{G} \right\} = \frac{G}{m^*(E)}, \tag{10.50}$$

which is just like eqn. (7.57) for parabolic bands but with the graphene effective mass given by eqn. (10.49). (To obtain this result, we did the differentiation at a constant energy, anticipating that later on, we'll integrate the constant energy result over all of the energy channels to get the total currents.)

If we continue and evaluate J_x and J_y, it's easy to believe that the result will be a magneto conductivity tensor like eqn. (7.84). The sign of the off-diagonal components will depend on the sign of the effective mass. The resulting Hall coefficient will be like the parabolic band case, eqn. (8.17); the sign will be negative for n-type conduction (Fermi level above the Dirac point) and positive for p-type conduction (Fermi level below the Dirac point). Working out the details is a useful exercise. (Hint: Assume $T = 0$ K to keep the math simple.)

Thermoelectric coefficients

So far, we have only discussed the conductivity in eqns. (10.1), but we can also work out the Seebeck coefficient and thermal conductivity of graphene. In this case, we cannot assume $T = 0$ K, or we won't get a finite answer, but the results are readily worked out in terms of Fermi Dirac integrals.

10.8 Summary

The purpose of this lecture has been to show that the concepts and approaches introduced in earlier lectures are quite general. Given a band structure, $E(k)$, one can readily work out expressions for the density-of-states, carrier density, number of conduction channels, conductivity, and the other parameters in the coupled current equations. Actually, the general model for transport as presented in Lecture 2 is even more general; it does not even require a crystal with a periodic structure that leads to a dispersion, $E(k)$. (See Datta's lecture notes for a discussion of this point [7].) In this lecture, we used graphene as an example of how to apply the concepts and formulas developed in previous lectures, but graphene is an interesting material in its own right, and the references provide some starting points for exploring this fascinating material.

10.9 References

Graphene is a two-dimensaional material of great scientific and technolofical interest. An introduction to its properties can be found at:

[1] "The Nobel Prize in Physics 2010". Nobelprize.org. 9 June 2011 http://nobelprize.org/nobel_prizes/physics/laureates/2010/

Datta gives an introduction to the band structure of graphene in two online lectures:

[2a] S. Datta, ECE 495N: Lecture 21, Fall 2008, https://nanohub.org/resources/5710

[2b] S. Datta, ECE 495N: Lecture 22, Fall 2008, https://nanohub.org/resources/5721

For a review of how to determine the density-of-states in k-space and in energy-space, see

[3] M.S. Lundstrom, "ECE 656: Lecture 2: Sums in k-space/Integrals in Energy Space, Fall 2009", http://nanohub.org/resources/7296

Consult these articles for information on the electronic properties of graphene:

[4a] S. das Sarma, S. Adam, E.H. Hwang, and E. Rossi, "Electronic transport in two-dimensinal graphene", *Rev. Modern Phys.*, **83**, pp. 407-470, 2011.

[4b] W. Zhu, V. Perebeinos, M. Freitag, and P. Avouris, "Carrier scattering, mobilities, and electrostatic potential in monolayer, bilayer, and trilayer graphene", *Phys. Rev. B.*, **80**, 235402, 2009.

[4c] V. Perebeinos and P. Avouris, "Inelastic scattering and current saturation in graphene", *Phys. Rev. B.*, **81**, 195442, 2010.

The notes below discuss near-equilibrium transport in graphene. They include a discussion of how to derive the scattering time for ADP scattering.

[5] Dionisis Berdebes, Tony Low, and Mark Lundstrom, "Lecture Notes on Low Bias Transport in Graphene: An Introduction," 2009. http:// nanohub.org/resources/7436/downloadNotes_on_low_field_transport_in_graphene.pdf

For a tutorial introduction to the near-equilibrium conductance of graphene, see:

[6] M.S. Lundstrom, "Colloquium on Graphene Physics and Devices: Lecture 3: Low Bias Transport in Graphene: An Introduction", http://nanohub.org/resources/7401

Datta discusses a general definition of the conductivity effective mass that works for parabolic energy bands as well as for graphene. See Lecture 5 in:

[7] Supriyo Datta, *Lessons from Nanoelectronics: A new approach to transport theory*, World Scientific Publishing Company, Singapore, 2011.

Appendix A

Summary of Key Results

The first part of this appendix is a summary of some key equations for near-equilibrium transport along with pointers to specifics in the notes. The goal is to collect in one place the equations needed for typical calculations. The second part of the appendix lists expressions for thermoelectric transport parameters in 1D, 2D, and 3D for parabolic band semiconductors with power law scattering. The last part of the appendix lists expressions for thermoelectric transport parameters of graphene, a 2D material with an unusual, but simple band structure.

General model for current

The general model for current in a nanodevice can be written in two, equivalent forms as given by eqn. (2.46):

$$I = \frac{2q}{h} \int \gamma(E)\pi \frac{D(E)}{2} (f_1 - f_2) \, dE$$

$$I = \frac{2q}{h} \int T(E)M(E) (f_1 - f_2) \, dE \,,$$

(A.1)

where γ is the broadening, defined in eqn. (2.13), and $D(E)$ is the density-of-states with the factor of 2 for spin included. In the second form, $M(E)$ is the number of channels (or modes) at energy, E. The transmission, $T(E)$, is given by eqn. (2.43) as

$$T(E) = \frac{\lambda(E)}{\lambda(E) + L} \,,$$

(A.2)

where λ is the mean-free-path for backscattering and L is the length of the conductor. General expressions for $M(E)$ are given in eqns. (2.25) or by eqns. (2.13) for parabolic bands and by eqn. (10.17) for graphene.

Near-equilibrium transport

For small applied bias (and constant temperature), we can expand $(f_1 - f_2)$ in eqn. (A.1) to find

$$(f_1 - f_2) \approx \left(-\frac{\partial f_0}{\partial E}\right) qV, \tag{A.3}$$

where $(-\partial f_0/\partial E)$ is known as the Fermi window function. Only channels where the magnitude of the window function is significant contribute to current flow. With this near-equilibrium assumption, we find the current, eqn. (2.50), to be

$$I = GV, \tag{A.4}$$

where the near-equilibrium conductance, G, is given by eqn. (2.51) as

$$G = \frac{2q^2}{h} \int T(E)M(E)\left(-\frac{\partial f_0}{\partial E}\right) dE. \tag{A.5}$$

For ballistic transport, $T(E) = 1$. For diffusive transport, $T(E) \approx \lambda(E)/L$, where L is the length of the sample. For a sample much longer than a mean-free-path, we obtain eqn. (2.56), the current equation for bulk (diffusive) transport as

$$J_{nx} = \sigma_n \frac{d(F_n/q)}{dx}, \tag{A.6}$$

where F_n is the electrochemical potential (also known as the quasi-Fermi level). The conductivity is given by eqn. (2.57) as

$$\sigma_n = \frac{2q^2}{h} \int M_{2D}(E)\lambda(E)\left(-\frac{\partial f_0}{\partial E}\right) dE, \tag{A.7}$$

We have assumed a 2D conductor in eqn. (A.7), but similar considerations apply in 1D and 3D. (Recall that $M_{2D}(E) = M(E)/W$.) Equations (3.56) show different, but mathematically equivalent ways of writing the conductivity.

Another way to write the conductance is as the product of the quantum of conductance, times the number of channels for conduction, times the average transmission as in eqn. (3.63)

$$G = \frac{2q^2}{h} \langle M \rangle \langle\langle T \rangle\rangle, \tag{A.8}$$

where (see eqns. (3.63))

$$\langle M \rangle \equiv \int M(E)\left(-\frac{\partial f_0}{\partial E}\right) dE, \tag{A.9}$$

and

$$\langle\langle T \rangle\rangle \equiv \frac{\int T(E)M(E)\left(-\frac{\partial f_0}{\partial E}\right)dE}{\int M(E)\left(-\frac{\partial f_0}{\partial E}\right)dE} = \frac{\langle MT \rangle}{\langle M \rangle}. \tag{A.10}$$

The quantity, $\langle M \rangle$ is simply the number of channels in the Fermi window.

Yet another way to write the conductance is in terms of the so-called differential conductance, $G'(E)$, as

$$G = \int G'(E)\,dE$$

$$G' = \frac{2q^2}{h}M(E)T(E)\left(-\frac{\partial f_0}{\partial E}\right). \tag{A.11}$$

Similar expressions apply for the conductivity, which can be written in terms of the differential conductivity as in eqn. (3.58) for 2D.

Temperature differences and gradients

When there are differences in both voltage and temperature across the device, then we must Taylor series expand $(f_1 - f_2)$ in both voltage and temperature to find, as in eqn. (5.6),

$$(f_1 - f_2) \approx \left(-\frac{\partial f_0}{\partial E}\right)q\Delta V - \left(-\frac{\partial f_0}{\partial E}\right)\frac{(E - E_F)}{T_L}\Delta T. \tag{A.12}$$

The result is an extra term in the current equation and also an equation for the heat current due to electrons:

$$I = G\Delta V + S_T\Delta T$$

$$I_Q = -T_L S_T \Delta V - K_0 \Delta T, \tag{A.13}$$

as in eqn. (5.24). In these equations, ΔV is the difference in voltage between contact 2 and contact 1, and ΔT is the difference in temperature. The current, I, is defined to be positive when it flows into contact 2 (electrons flowing out of contact 2). The heat current, I_Q, is positive when it flows in the $+x$ direction — out of contact 2. The four transport coefficients in eqn. (A.13) are given by eqn. (5.25) as

$$G'(E) = \frac{2q^2}{h}T(E)M(E)\left(-\frac{\partial f_0}{\partial E}\right)$$

$$G = \int G'(E)dE$$

$$S_T = -\left(\frac{k_B}{q}\right)\int\left(\frac{E - E_F}{k_B T_L}\right)G'(E)dE$$

$$K_0 = T_L\left(\frac{k_B}{q}\right)^2\int\left(\frac{E - E_F}{k_B T_L}\right)^2 G'(E)dE.$$

(A.14)

For long, diffusive samples, we can write eqns. (A.13) in the common form used to describe bulk transport as

$$J_{nx} = \sigma_n\frac{d(F_n/q)}{dx} - s_T\frac{dT_L}{dx}$$

$$J_{Qx} = T_L s_T\frac{d(F_n/q)}{dx} - \kappa_0\frac{dT_L}{dx},$$

(A.15)

which is eqn. (5.26). The four transport coefficients are given by eqn. (5.27) as

$$\sigma'_n(E) = \frac{2q^2}{h}M_{3D}(E)\lambda(E)\left(-\frac{\partial f_0}{\partial E}\right)$$

$$\sigma_n = \int \sigma'_n(E)dE$$

$$s_T = -\left(\frac{k_B}{q}\right)\int\left(\frac{E - E_F}{k_B T_L}\right)\sigma'_n(E)dE$$

$$\kappa_0 = T_L\left(\frac{k_B}{q}\right)^2\int\left(\frac{E - E_F}{k_B T_L}\right)^2 \sigma'_n(E)dE.$$

(A.16)

These equations are written for 3D conductors (recall that $M_{3D}(E) = M(E)/A$). Similar equations can be written for 2D and 1D transport.

In practice, the inverted form of these equations is often preferred. Using eqn. (5.28), the inverted form of eqn. (A.13) becomes

$$\Delta V = RI - S\Delta T$$

$$I_Q = -\Pi I - K_n\Delta T,$$

(A.17)

where

$$S = \frac{S_T}{G}$$

$$\Pi = T_L S \tag{A.18}$$

$$K_n = K_0 - \Pi S G .$$

Similarly, the inverted form of the bulk transport equations become, eqn. (5.30),

$$\frac{d\left(F_n/q\right)}{dx} = \rho_n J_{nx} + S_n \frac{dT_L}{dx}$$

$$J_{Qx} = T_L S_n J_{nx} - \kappa_n \frac{dT_L}{dx} , \tag{A.19}$$

which should be compared to eqns. (5.26). The transport parameters in eqns. (5.30) are

$$\rho_n = 1/\sigma_n$$

$$S_n = s_T/\sigma_n \tag{A.20}$$

$$\kappa_n = \kappa_0 - S_n^2 \sigma_n T_L .$$

In summary, given a band structure, the number of channels, $M(E)$, can be evaluated from eqns. (2.25) and, if a model for the mean-free-path for backscattering, $\lambda(E)$ can be obtained, then the near-equilibrium transport parameters can be evaluated using the expressions listed above. For parabolic energy bands and power law scattering, analytical expressions in terms of Fermi-Dirac integrals can be obtained.

Thermoelectric coefficients for parabolic band semiconductors in 1D, 2D, and 3D

In the expressions for the various transport parameters listed below, parabolic energy bands:

$$E(k) = \frac{\hbar^2 k^2}{2m^*} , \tag{A.21}$$

and power law scattering:

$$\lambda(E) = \lambda_0 \left(\frac{E}{k_B T_L}\right)^r, \tag{A.22}$$

are assumed. Unipolar (electron) conduction is assumed; bipolar conduction can be treated as discussed in Sec. 5.5. The location of the Fermi level in relation to the band edge is described by the dimensionless parameter:

$$\eta_F = \frac{(E_F - E_C)}{k_B T_L}, \tag{A.23}$$

where E_C is the conduction band edge.

The results listed below are taken from Appendix B. of the Ph.D. thesis of Raseong Kim, *Physics and Simulation of Nanoscale Electronics and Thermoelectric Devices*, Purdue University, West Lafayette, Indiana, U.S.A., August, 2011.

Thermoelectric coefficients in 1D: Ballistic

$$G = \frac{2q^2}{h} \mathcal{F}_{-1}(\eta_F)$$

$$S_T = -\frac{k_B}{q} \frac{2q^2}{h} \left(\mathcal{F}_0(\eta_F) - \eta_F \mathcal{F}_{-1}(\eta_F)\right)$$

$$K_0 = T_L \left(\frac{k_B}{q}\right)^2 \frac{2q^2}{h} \left(2\mathcal{F}_1(\eta_F) - 2\eta_F \mathcal{F}_0(\eta_F) + \eta_F^2 \mathcal{F}_{-1}(\eta_F)\right) \tag{A.24}$$

$$S = -\frac{k_B}{q} \left(\frac{\mathcal{F}_0(\eta_F)}{\mathcal{F}_{-1}(\eta_F)} - \eta_F\right)$$

$$K_n = T_L \left(\frac{k_B}{q}\right)^2 \frac{2q^2}{h} \left(2\mathcal{F}_1(\eta_F) - \frac{\mathcal{F}_0^2(\eta_F)}{\mathcal{F}_{-1}(\eta_F)}\right).$$

Thermoelectric coefficients in 1D: Diffusive

$$G = \frac{2q^2}{h} \left(\frac{\lambda_0}{L} \right) \Gamma(r+1) \mathcal{F}_{r-1}(\eta_F)$$

$$S_T = -\frac{k_B}{q} \frac{2q^2}{h} \left(\frac{\lambda_0}{L} \right) \Gamma(r+1) \left((r+1)\mathcal{F}_r(\eta_F) - \eta_F \mathcal{F}_{r-1}(\eta_F) \right)$$

$$K_0 = T_L \left(\frac{k_B}{q} \right)^2 \frac{2q^2}{h} \left(\frac{\lambda_0}{L} \right)$$

$$\times \left(\Gamma(r+3)\mathcal{F}_{r+1}(\eta_F) - 2\eta_F \Gamma(r+2)\mathcal{F}_r(\eta_F) + \eta_F{}^2 \Gamma(r+1)\mathcal{F}_{r-1}(\eta_F) \right)$$

$$S = -\frac{k_B}{q} \left(\frac{(r+1)\mathcal{F}_r(\eta_F)}{\mathcal{F}_{r-1}(\eta_F)} - \eta_F \right)$$

$$K_n = T_L \left(\frac{k_B}{q} \right)^2 \frac{2q^2}{h} \left(\frac{\lambda_0}{L} \right) \Gamma(r+2) \left((r+2)\mathcal{F}_{r+1}(\eta_F) \right.$$

$$\left. - \frac{(r+1)\mathcal{F}_r^2(\eta_F)}{\mathcal{F}_{r-1}(\eta_F)} \right).$$

$$(A.25)$$

To find the 1D conductivity in siemen-meters (S-m) as opposed to the conductance in seimens as listed above, recall that

$$G = \sigma_{1D} \left(\frac{1}{L} \right), \tag{A.26}$$

so

$$\sigma_{1D} = GL. \tag{A.27}$$

Similarly, we have

$$s_T = S_T L$$
$$\kappa_0 = K_0 L \tag{A.28}$$
$$\kappa_n = K_n L.$$

Thermoelectric coefficients in 2D: Ballistic

$$G = W \frac{2q^2}{h} \left(\frac{\sqrt{2\pi m^* k_B T_L}}{h} \right) \mathcal{F}_{-1/2}(\eta_F)$$

$$S_T = -W \frac{k_B}{q} \frac{2q^2}{h} \left(\frac{\sqrt{2\pi m^* k_B T_L}}{h} \right) \left(\frac{3}{2} \mathcal{F}_{1/2}(\eta_F) - \eta_F \mathcal{F}_{-1/2}(\eta_F) \right)$$

$$K_0 = W T_L \left(\frac{k_B}{q} \right)^2 \frac{2q^2}{h} \left(\frac{\sqrt{2\pi m^* k_B T_L}}{h} \right)$$

$$\times \left(\frac{15}{4} \mathcal{F}_{3/2}(\eta_F) - 3\eta_F \mathcal{F}_{1/2}(\eta_F) + \eta_F{}^2 \mathcal{F}_{-1/2}(\eta_F) \right)$$

$$S = -\frac{k_B}{q} \left(\frac{3\mathcal{F}_{1/2}(\eta_F)}{2\mathcal{F}_{-1/2}(\eta_F)} - \eta_F \right)$$

$$K_n = W T_L \left(\frac{k_B}{q} \right)^2 \frac{2q^2}{h} \left(\frac{\sqrt{2\pi m^* k_B T_L}}{h} \right) \left(\frac{15}{4} \mathcal{F}_{3/2}(\eta_F) - \frac{9\mathcal{F}_{1/2}^2(\eta_F)}{4\mathcal{F}_{-1/2}(\eta_F)} \right).$$

$$\text{(A.29)}$$

Thermoelectric coefficients in 2D: Diffusive

$$G = W \frac{2q^2}{h} \left(\frac{\lambda_0}{L} \right) \left(\frac{\sqrt{2m^* k_B T_L}}{\pi \hbar} \right) \Gamma \left(r + \frac{3}{2} \right) \mathcal{F}_{r-1/2}(\eta_F)$$

$$S_T = -W \frac{k_B}{q} \frac{2q^2}{h} \left(\frac{\lambda_0}{L} \right) \left(\frac{\sqrt{2m^* k_B T_L}}{\pi \hbar} \right)$$

$$\times \left(\Gamma \left(r + \frac{5}{2} \right) \mathcal{F}_{r+1/2}(\eta_F) - \eta_F \Gamma \left(r + \frac{3}{2} \right) \mathcal{F}_{r-1/2}(\eta_F) \right)$$

$$K_0 = W T_L \left(\frac{k_B}{q} \right)^2 \frac{2q^2}{h} \left(\frac{\lambda_0}{L} \right) \left(\frac{\sqrt{2m^* k_B T_L}}{\pi \hbar} \right)$$

$$\times \left(\Gamma \left(r + \frac{7}{2} \right) \mathcal{F}_{r+3/2}(\eta_F) - 2\eta_F \Gamma \left(r + \frac{5}{2} \right) \mathcal{F}_{r+1/2}(\eta_F) \right.$$

$$\left. + \eta_F{}^2 \Gamma \left(r + \frac{3}{2} \right) \mathcal{F}_{-1/2}(\eta_F) \right)$$

$$S = -\frac{k_B}{q}\left(\frac{(r+3/2)\mathcal{F}_{r+1/2}(\eta_F)}{\mathcal{F}_{r-1/2}(\eta_F)} - \eta_F\right)$$

$$K_n = WT_L\left(\frac{k_B}{q}\right)^2\frac{2q^2}{h}\left(\frac{\lambda_0}{L}\right)\left(\frac{\sqrt{2m^*k_BT_L}}{\pi\hbar}\right)\Gamma\left(r+\frac{5}{2}\right)$$

$$\times\left(\left(r+\frac{5}{2}\right)\mathcal{F}_{r+3/2}(\eta_F) - \frac{(r+\frac{3}{2})\mathcal{F}_{r+1/2}^2(\eta_F)}{\mathcal{F}_{r-1/2}(\eta_F)}\right).$$

$$(A.30)$$

To find the 2D conductivity in siemens (S) (or S/square) as opposed to the conductance in siemens as listed above, recall that

$$G = \sigma_{2D}\left(\frac{W}{L}\right), \tag{A.31}$$

so

$$\sigma_{2D} = G\left(\frac{L}{W}\right). \tag{A.32}$$

Similar expressions apply for the other transport coefficients.

Thermoelectric coefficients in 3D: Ballistic

$$G = A\frac{2q^2}{h}\left(\frac{m^*k_BT_L}{2\pi\hbar^2}\right)\mathcal{F}_0(\eta_F)$$

$$S_T = -A\frac{k_B}{q}\frac{2q^2}{h}\left(\frac{m^*k_BT_L}{2\pi\hbar^2}\right)(2\mathcal{J}_1(\eta_F) - \eta_F\mathcal{F}_0(\eta_F))$$

$$K_0 = AT_L\left(\frac{k_B}{q}\right)^2\frac{2q^2}{h}\left(\frac{m^*k_BT_L}{2\pi\hbar^2}\right)$$

$$\times\left(6\mathcal{F}_2(\eta_F) - 4\eta_F\mathcal{F}_1(\eta_F) + \eta_F^2\mathcal{F}_0(\eta_F)\right)$$

$$(A.33)$$

$$S = -\frac{k_B}{q}\left(\frac{2\mathcal{F}_1(\eta_F)}{\mathcal{F}_0(\eta_F)} - \eta_F\right)$$

$$K_n = AT_L\left(\frac{k_B}{q}\right)^2\frac{2q^2}{h}\left(\frac{m^*k_BT_L}{2\pi\hbar^2}\right)\left(6\mathcal{F}_2(\eta_F) - \frac{4\mathcal{F}_1^2(\eta_F)}{\mathcal{F}_0(\eta_F)}\right).$$

Thermoelectric coefficients in 3D: Diffusive

$$G = A\frac{2q^2}{h}\left(\frac{\lambda_0}{L}\right)\left(\frac{m^*k_BT_L}{2\pi\hbar^2}\right)\Gamma(r+2)\mathcal{F}_r(\eta_F)$$

$$S_T = -A\frac{k_B}{q}\frac{2q^2}{h}\left(\frac{\lambda_0}{L}\right)\left(\frac{m^*k_BT_L}{2\pi\hbar^2}\right)$$

$$\times\left(\Gamma(r+3)\mathcal{F}_{r+1}(\eta_F)-\eta_F\Gamma(r+2)\mathcal{F}_r(\eta_F)\right)$$

$$K_0 = AT_L\left(\frac{k_B}{q}\right)^2\frac{2q^2}{h}\left(\frac{\lambda_0}{L}\right)\left(\frac{m^*k_BT_L}{2\pi\hbar^2}\right)$$

$$\times\left(\Gamma(r+4)\mathcal{F}_{r+2}(\eta_F)-2\eta_F\Gamma(r+3)\mathcal{F}_{r+1}(\eta_F)\right.\tag{A.34}$$

$$\left.+\eta_F{}^2\Gamma(r+2)\mathcal{F}_r(\eta_F)\right)$$

$$S = -\frac{k_B}{q}\left(\frac{(r+2)\mathcal{F}_{r+1}(\eta_F)}{\mathcal{F}_r(\eta_F)}-\eta_F\right)$$

$$K_n = AT_L\left(\frac{k_B}{q}\right)^2\frac{2q^2}{h}\left(\frac{\lambda_0}{L}\right)\left(\frac{m^*k_BT_L}{2\pi\hbar^2}\right)$$

$$\times\Gamma(r+3)\left((r+3)\mathcal{F}_{r+2}(\eta_F)-\frac{(r+2)\mathcal{F}_{r+1}^2(\eta_F)}{\mathcal{F}_r(\eta_F)}\right).$$

To find the 3D conductivity in siemens/meter (S/m) as opposed to the conductance in siemens as listed above, recall that

$$G = \sigma_{3D}\left(\frac{A}{L}\right),\tag{A.35}$$

so

$$\sigma_{3D} = G\left(\frac{L}{A}\right).\tag{A.36}$$

Similar expressions apply for the other transport coefficients.

Thermoelectric coefficients for graphene

Graphene is a two-dimensional material with a unique band structure,

$$E(k) = \hbar v_F k = \hbar v_F\sqrt{k_x^2+k_y^2}.\tag{A.37}$$

The transport parameters can be evaluated from eqns. (A.14) and (A.18) with the number of channels given by eqn. (10.17) as

$$M(E) = \frac{2|E|}{\pi \hbar v_F} W.$$

(A.38)

For scattering described in power law form as in eqn. (A.22), we can evaluate the thermoelectric parameters for graphene in terms of Fermi-Dirac integrals. Since the bandgap is zero, bipolar conduction should be considered whenever $T_L > 0$ K.

The results listed below were provided by Dr. Raseong Kim.

Thermoelectric coefficients of graphene: Ballistic

$$G = W \frac{2q^2}{h} \left(\frac{2k_B T_L}{\pi \hbar v_F} \right) (\mathcal{F}_0(\eta_F) + \mathcal{F}_0(-\eta_F))$$

$$S_T = -W \left(\frac{k_B}{q} \right) \frac{2q^2}{h} \left(\frac{2k_B T_L}{\pi \hbar v_F} \right) [2(\mathcal{F}_1(\eta_F) - \mathcal{F}_1(-\eta_F))$$
$$- \eta_F (\mathcal{F}_0(\eta_F) + \mathcal{F}_0(-\eta_F))]$$

$$K_0 = W T_L \left(\frac{k_B}{q} \right)^2 \frac{2q^2}{h} \left(\frac{2k_B T_L}{\pi \hbar v_F} \right)$$
$$\times [6(\mathcal{F}_2(\eta_F) + \mathcal{F}_2(-\eta_F)) - 4\eta_F (\mathcal{F}_1(\eta_F) - \mathcal{F}_1(-\eta_F))$$
$$+ \eta_F{}^2 (\mathcal{F}_0(\eta_F) + \mathcal{F}_0(-\eta_F))]$$

(A.39)

$$S = - \left(\frac{k_B}{q} \right) \left(\frac{2(\mathcal{F}_1(\eta_F) - \mathcal{F}_1(-\eta_F))}{\mathcal{F}_0(\eta_F) + \mathcal{F}_0(-\eta_F)} - \eta_F \right)$$

$$K_n = W T_L \left(\frac{k_B}{q} \right)^2 \frac{2q^2}{h} \left(\frac{2k_B T_L}{\pi \hbar v_F} \right)$$
$$\times \left(6(\mathcal{F}_2(\eta_F) + \mathcal{F}_2(-\eta_F)) - \frac{4(\mathcal{F}_1(\eta_F) - \mathcal{F}_1(-\eta_F))^2}{\mathcal{F}_0(\eta_F) + \mathcal{F}_0(-\eta_F)} \right).$$

Thermoelectric coefficients of graphene: Diffusive

$$G = W \frac{2q^2}{h} \left(\frac{\lambda_0}{L}\right) \left(\frac{2k_B T_L}{\pi \hbar v_F}\right) \Gamma(r+2)\big(\mathcal{F}_r(\eta_F) + \mathcal{F}_r(-\eta_F)\big)$$

$$S_T = -W \left(\frac{k_B}{q}\right) \frac{2q^2}{h} \left(\frac{\lambda_0}{L}\right) \left(\frac{2k_B T_L}{\pi \hbar v_F}\right)$$

$$\times \big[\Gamma(r+3)\big(\mathcal{F}_{r+1}(\eta_F) - \mathcal{F}_{r+1}(-\eta_F)\big)$$

$$- \eta_F \Gamma(r+2)\big(\mathcal{F}_r(\eta_F) + \mathcal{F}_r(-\eta_F)\big)\big]$$

$$K_0 = W T_L \left(\frac{k_B}{q}\right)^2 \frac{2q^2}{h} \left(\frac{\lambda_0}{L}\right) \left(\frac{2k_B T_L}{\pi \hbar v_F}\right)$$

$$\times \big[\Gamma(r+4)\big(\mathcal{F}_{r+2}(\eta_F) + \mathcal{F}_{r+2}(-\eta_F)\big) - 2\eta_F \Gamma(r+3)\big(\mathcal{F}_{r+1}(\eta_F)$$

$$- \mathcal{F}_{r+1}(-\eta_F)\big) + \eta_F^2 \Gamma(r+2)\big(\mathcal{F}_r(\eta_F) + \mathcal{F}_r(-\eta_F)\big)\big]$$

$$S = -\left(\frac{k_B}{q}\right) \left(\frac{(r+2)\big(\mathcal{F}_{r+1}(\eta_F) - \mathcal{F}_{r+1}(-\eta_F)\big)}{\mathcal{F}_r(\eta_F) + \mathcal{F}_r(-\eta_F)} - \eta_F\right)$$

$$K_n = W T_L \left(\frac{k_B}{q}\right)^2 \frac{2q^2}{h} \left(\frac{\lambda_0}{L}\right) \left(\frac{2k_B T_L}{\pi \hbar v_F}\right)$$

$$\times \Gamma(r+3)\bigg((r+3)\big(\mathcal{F}_{r+2}(\eta_F) + \mathcal{F}_{r+2}(-\eta_F)\big)$$

$$- \frac{(r+2)\big(\mathcal{F}_{r+1}(\eta_F) - \mathcal{F}_{r+1}(-\eta_F)\big)^2}{\mathcal{F}_r(\eta_F) + \mathcal{F}_r(-\eta_F)}\bigg).$$

$$\text{(A.40)}$$

To find the 2D conductivity of graphene in siemens (S) (or S/square) as opposed to the conductance in siemens as listed above, refer to eqns. (A.31)–(A.32).

Index